Riccardo Levi-Setti's photographic study of trilobites, with his accompanying textual analysis, illustrates an unusual collection of beautifully preserved trilobite specimens. Aesthetically pleasing and scientifically valuable, these dramatic black and white photographs are the result of Levi-Setti's inventive photographic techniques. The author has adapted his techniques of macrophotography to the special configuration of each example of these ancient fossils.

Produced in large format, these high contrast black-and-white prints are uniquely suited to reveal rich details of complex structures which are otherwise difficult or impossible to detect on most museum specimens. Complementing the prints is a clear and complete account of the nature and life of trilobites.

Trilobites, the author observes, are not only fascinating, but also extremely important fossils. These animals first appeared in the early Cambrian period, about 600 million years ago, and lived for 350 million years. Their fossilized remains are found throughout the world; they are invaluable both as geologic indices and as clues to the evolution of the first complex life-forms on earth.

Professor Levi-Setti's study includes an account of the discovery of remarkable sophistication in the trilobites' optical organ. While attempting a physical interpretation of trilobite lens structures reported by Euan N. K. Clarkson, the author found that they corresponded to the design of optically corrected lenses once suggested by Descartes and Huygens. Just before publication, on a trip to Newfoundland, Levi-Setti made another discovery: giant trilobites, of a species as yet unreported from the North American continent, in fact identical to the first Swedish trilobite described by Linnaeus two hundred years ago. A special appendix incorporates these findings.

As Euan N. K. Clarkson says in the foreword, "There has long been a need for such a book as this. Within it there is an unrivaled pictorial record of some of the most remarkable animals ever to have lived on earth."

Riccardo Levi-Setti received his doctoral degree from the University of Pavia in 1949. He came to the United States in 1956 and is professor of physics at the University of Chicago. He has had a long-standing acquaintance with trilobites.

Trilobites

Trilobites

A Photographic Atlas

Riccardo Levi-Setti

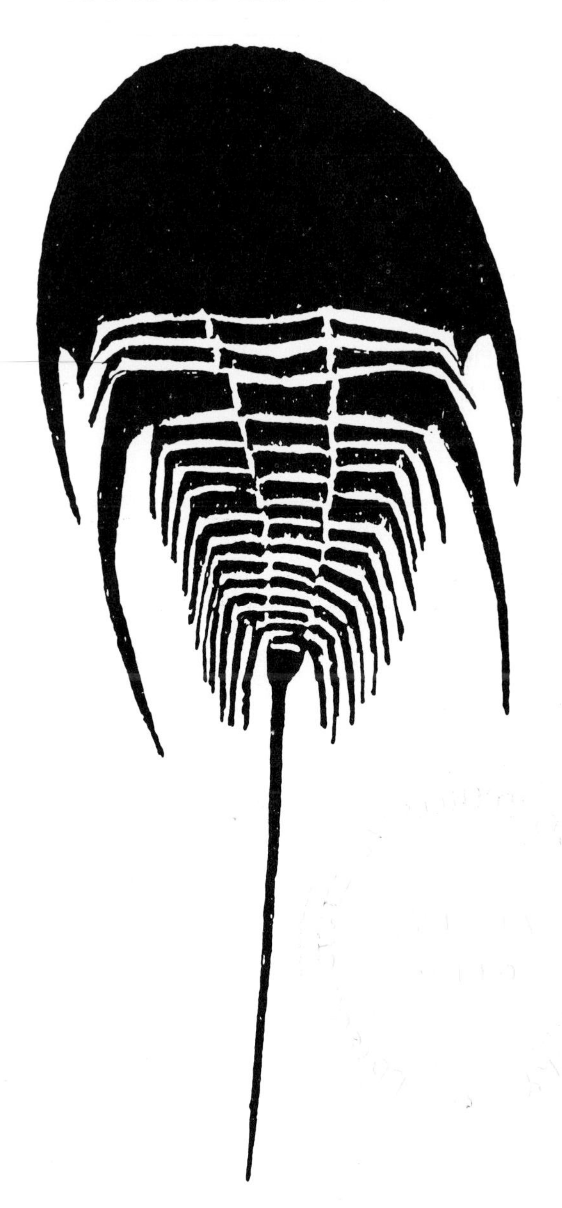

University of Chicago Press

Chicago and London

The University of Chicago Press,
Chicago 60637
The University of Chicago Press,
Ltd., London

Library of Congress Cataloging in Publication Data

Levi-Setti, Riccardo.
Trilobites: a photographic atlas.

Bibliography: p.
1. Trilobites—Pictorial works. I. Title.
QE821.L47 565'.393 74-7555
ISBN 0-226-47448-8

Contents

Foreword

There is a perennial fascination in the study of trilobites for professional scholars and amateur natural historians alike. The unique form and vast antiquity of these ancient fossils compel our immediate attention.

Trilobites lived in the Palaeozoic oceans for some 350 million years of geological time. During their immensely long history they evolved into diverse forms and colonized numerous environments. They became extinct at the end of Permian time, over 200 million years ago. A great deal is now known about trilobites. They have long been valued by geologists as stratigraphic indicators, and their basic anatomy, their growth and development from the larval stages, the nature of their appendages, cuticular structure, and sense organs, their evolutionary differentiation, and their distribution in time and space have all been the subject of intensive research. But the virtual absence of preserved soft parts imposes strict limitations upon what can be known, and certain important matters may remain forever cryptic, and even the fundamentals of classification are still disputed.

The would-be collector and student of trilobites is often limited in his endeavors by the paucity of really well preserved material for study. One does not often find perfect specimens. These occur only in certain rock-types in specific locales, some of which are no longer accessible. They may be difficult to extract from the matrix, and even in the best displayed museum specimens it is not easy to see the microscopic details of structure. In this book these difficulties are largely overcome. Professor Levi-Setti, a physicist with a long-standing acquaintance with trilobites, has performed a timely service in setting out a clear account of the nature and life of trilobites and providing a systematic atlas of superb photographs. He has selected the most elegantly preserved trilobites of all which display in an unprecedented manner the rich details of form and structure of these remarkable organisms. The photographs, which are nearly all Professor Levi-Setti's own, are the result of years of careful and painstaking work. They are an aesthetic delight in themselves but more importantly they show the high degree of biological organization and adaptational complexity which these extinct animals attained as long ago as the remote Palaeozoic. In no way were trilobites low-grade organisms. Indeed, their structure, within its own phylogenetic limits, compares favorably with that of many modern marine arthropods.

There has long been a need for such a book as this. Within it there is an unrivaled pictorial record of some of the most remarkable animals ever to have lived on earth. Though long since extinct the trilobites retain the power to captivate.

Euan N. K. Clarkson
Edinburgh

Acknowledgments

I wish to express my gratitude to a great number of persons who over the years have contributed encouragement, inspiration, and practical help. I am particularly indebted to Drs. Eugene Richardson and Matthew Nitecki of the Field Museum of Natural History, Chicago, for making the museum collections available for the selection of specimens to be used in this work, for the loan of those specimens, and for many helpful discussions. I wish to thank Dr. Bernhard Kummel of Harvard's Museum of Comparative Zoology for allowing my perusal of the museum collection, and for the contribution of some of the most beautiful trilobites in the entire Atlas, from Bohemia in particular. The section on trilobite eyes would not have existed without the contributions by Dr. E. N. K. Clarkson of the Grant Institute of Geology, Edinburgh. For his magnificent photographs and the communication of many of his results a great deal of appreciation is due. Through the courtesy of Dr. J. Cisne, I had access to the collection of Beecher's Utica shale trilobites as described in section 3.2. In addition to the loan of many of the trilobites, many x-ray negatives were most generously contributed by Dr. Cisne, to whom I am deeply grateful. Dr. J. Bergström, of the University of Lund, has been most helpful in discussing questions of classification. I am also grateful to Dr. N. Eldredge of the American Museum of Natural History and to Dr. K. Towe of the Smithsonian Institution for their discussions and useful communications of their work. Dr. A. M. Zeigler at the University of Chicago has helped on many occasions with his criticism and stimulating discussions.

Further appreciation is due the superintendents of many quarries for their permission to collect fossils; particular thanks go to Mr. H. Nester of the Consumer's Company Quarry, at McCook, Illinois; Mr. J. Riordan of the Lehigh Stone Company, Lehigh, Illinois; and the Medusa Cement Company at Sylvania, Ohio. Many more persons have helped—with permission to collect, with loans of specimens, and with precious gifts. Among the latter is Mr. M. Gazay of the International Red Cross Organization, Geneva, Switzerland, who has donated several beautiful trilobites. A word of appreciation is also due a number of fossil dealers, from whom many specimens have been obtained. Among these, Mr. Afton Fawcett of Hurricane, Utah, has provided exceptional trilobite specimens, collected and prepared with professional care. I am particularly indebted to my sons Emile and Matteo, who since they were first able to climb boulders have very successfully competed with their father in finding countless trilobites.

Once, several years ago, in Rome, I had the opportunity of meeting Franco Rasetti. The memory of that meeting is still cherished, and those who know of Rasetti's contributions to physics and paleontology will understand the reasons for a particular devotion to this great scientist on the part of another trilobite-loving physicist.

Finally, the careful work of Gail Goldblatt in deciphering my handwriting and typing this manuscript is sincerely appreciated.

Introduction 1

Among the remains of early life on earth, the fossil record which we find buried in ancient sedimentary rocks bears evidence of an extraordinary group of marine creatures, the trilobites. The position of these invertebrates in the evolution of the animal kingdom is extraordinary because they were the first highly organized animals to populate the primordial seas. Trilobites could see their immediate environment with amazingly sophisticated optical devices in the form of large composite eyes, the first use of optics coupled with sensory perception in nature. As a unique feat in the history of life, their eye lenses were shaped to correct for optical aberrations, with design identical to that proposed (quite independently of any knowledge of trilobites) by Descartes and Huygens.

Although we can only hold the petrified remains of this long-extinct form of life, what was preserved of trilobites constitutes a record of immediate and striking impact, a still life which we can interpret and recognize. It is this extravagantly rich still life which we want to illustrate in this book, preserving as much as possible of the excitement of a voyage back in time to the dawn of life.

The true story of trilobites must have begun in a very nebulous and ancient period of the history of the earth, vaguely defined as the Precambrian period, perhaps as far back in time as one billion years. We have no record of this beginning; however, we infer that a tremendous evolutionary process must have occurred prior to the first recorded occurrence of trilobites in sedimentary rocks. In fact, trilobites appear suddenly in the fossil record at the very beginning of the Cambrian period, some 600 million years ago, when they had already reached a full degree of development and differentiation. Compound eyes are already present in the earliest trilobites. The preservation of trilobites as fossils is related to the presence of a calcified or otherwise mineralized exoskeleton in the living animal. It is as if the shells of trilobites became rich in calcium carbonate or phosphate rather abruptly in the course of evolution, and it is only after that time that they could be preserved in their burial rock. The high point of the trilobites' differentiation of forms was reached toward the end of the Cambrian period, some 500 million years ago. They became extinct at the end of the Permian period, the borderline between the Paleozoic and Mesozoic eras, around 230 million years ago. Dinosaurs were still in the making at that time, to appear some 50 million years later. Long before the age of fishes, which began about 400 million years ago, the trilobites had already "ruled the seas" (there was very little else to rule upon) for several hundred million years.

The span of time which saw the birth, development, and disappearance of trilobites and their age is indeed staggering. However, the fact that their general appearance is not so dissimilar from that of living arthropods—the horseshoe crab, for example—makes trilobites less forbidding than some other forms of extinct life.

The widespread abundance of trilobites, particularly during the Cambrian period, is evident from layers of sedimentary rocks which are occasionally coated with trilobite remains. The presence of particular genera in layers of determinate geologic age makes trilobites very important index fossils. The presence of identical forms in rocks of identical age and composition on locations which are now separated by oceans is telltale evidence of the drift of the continents on the earth's crust.

The fascination of the trilobites' age, their role in early life on our planet, the optimization of their visual organs, the ingenuity of their life adaptation to the environment, and their value as geological markers explain why trilobites are important to both the professional scientist and the amateur fossil collector. The photographs in this book are meant for both groups of trilobite enthusiasts. Perhaps the varied forms and meaning of these creatures may awake the interest and excite the imagination of a broader group of readers. It is hoped that these pictures may reach all those who still marvel at the wonders of nature, the remarkable forms of present life as well as those no less remarkable of a very, very distant past.

Fig. 1. Composition of the animal kingdom through geologic time. Only the principal phyla are indicated. The time scale indicated was adopted by the Geological Society of London in 1964 (from Kummel 1970). The histogram at the bottom of the figure indicates the percentage of living species, excluding insects (partially from Easton 1960; Shrock and Twenhofel 1953; and from a survey by the author).

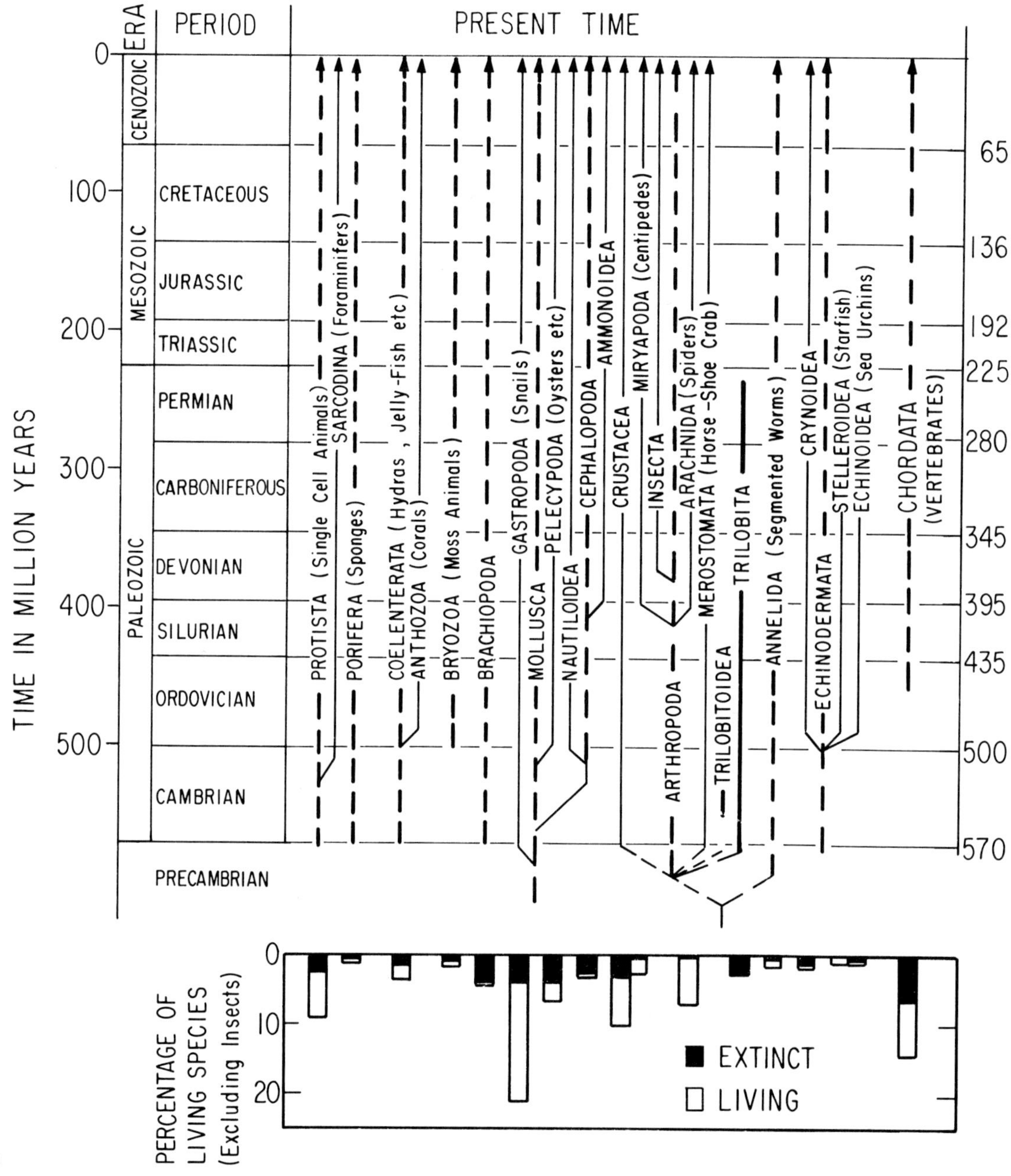

Fig. 1

The Arthropoda 2

In the subdivision of the animal kingdom into major groups or phyla, the trilobites are included in the phylum Arthropoda. The history of these highly developed invertebrates reaches back into the Precambrian and extends to modern times. The oldest fossil arthropods are trilobites. Familiar arthropods are the insects, scorpions, spiders, centipedes and millipedes, shrimps, lobsters, crabs, and others with jointed legs. Although they first inhabited the seas, the arthropods have adapted themselves most efficiently to every life habitat.

The position of the Arthropoda, and of the Trilobita in particular, in the animal kingdom is illustrated in figure 1. Here the principal phyla, represented by vertical broken lines, are traced through the Phanerozoic time scale (the ensemble of geologic eras with a fossil record), from their first appearance to present times. Some of the geologically important further subdivisions of the phyla are indicated as full lines. This representation is very schematic and does not show the variations in the population of each phylum as a function of time. In the histogram at the base of figure 1, however, we can appreciate a cumulative comparison of the relative abundance of known extinct and living species. The Arthropoda actually represent a much more prolific group of animals than is shown in such a histogram. In fact, the 1 million or more species of insects, representing 70 percent of all species of extinct and living animals, had to be removed from the plot so that the remnant fractions could become visible. This may come as a surprise to many, but from the point of view of a statistician of living species, we live today in the age of insects! The early Paleozoic was the age of trilobites.

Basically the arthropods have segmented bodies, of elongated form, which exhibit bilateral symmetry about an axis carrying the digestive tract. The soft body is encased in a protective shield, or *exoskeleton,* made of chitin, which may be hardened by calcium carbonate or phosphate. The segments, or *somites*, serve the function of articulating the body and usually carry pairs of appendages which are, in turn, divided into jointed segments (hence the name Arthropoda, from the Greek *arthron* = joint, *podos* = foot). The appendages may become differentiated (e.g., into antennulae, claws, etc.) to serve a variety of specialized functions. Groups of somites are often differentiated also and can be fused together. Thus one can usually recognize *head, thorax,* and *abdomen*. The somites of the head are fused together in all arthropods. As we shall see, other groups of somites may become fused. In extreme cases, as in the horseshoe crab (*Limulus*), fusion occurs between head and thorax to form a solid shield called *cephalo-thorax*.

The nervous system of the arthropods is highly developed, with pairs of ganglia in most somites, and a pair of nerve cords leading to a more or less developed brain. The interaction of the animal with the environment takes place by means of sensory organs, such as tactile antennulae and single or compound eyes. The latter, as we shall see in detail for the trilobites, can attain surprising complexity and optimization of function. Arthropods possess a circulatory system consisting of heart, arteries, and blood return ducts. Respiration takes place through *gills* in the aquatic forms or through a network of *tracheae* in the air-breathing forms. The sexes are distinct, and eggs hatch either externally or internally. Growth takes many forms. We are familiar with the *metamorphosis* of the caterpillar into a moth or butterfly. The presence of a hard, inextensible exoskeleton poses special problems to the growth of many arthropods. In such cases the increase in size takes place by *ecdysis* (*molting*), in which the hard cover is shed periodically, and a new, larger shield is constructed after each step. Soft-shell crabs are examples of the stage of growth following ecdysis. For a period of time the animal is protected only by a flexible, chitinous shell which will gradually harden by mineralization.

In figure 1 the phylum Arthropoda is shown to be connected in some way to the Annelida back in Precambrian times. Several features do in fact exist in common between the two groups of animals, in particular the body segmentation, the structure, and organization of the nervous system. The existence of a primordial, annelid-like ancestor of both Annelida and Arthropoda is therefore postulated, although no fossil record exists for these proto-Annelida.
In any case, the Arthropoda, and in particular the Trilobita, appear in the fossil record at an already advanced stage of evolution and differentiation.

3 The Trilobita

3.1. Introduction to Trilobite Morphology

The general description of the distinctive features of the Arthropoda given in chapter 2 applies to the Trilobita as well. Here, however, the picture is less complete than for the living arthropods, since, of course, our evidence must be based only on the fossil record. Several details of the anatomy and physiology of trilobites are inferred by analogy with what is known about the character of arthropods in general.

Trilobites were, on the average, small animals, two to seven centimeters long, with extremes ranging from three millimeters to about seventy centimeters. They lived in the sea, for their remains are always associated with those of marine animals (corals, brachiopods, cephalopods, and so forth).

The commonly preserved portion of the body of trilobites is the dorsal shield, or *carapace,* made of mineralized chitin. The ventral part of the exoskeleton is never preserved, and we must infer that it consisted of a membrane made of chitin or other nonmineralized substance. As a rule, all soft parts as well as appendages are also missing in the fossil trilobite. There are, however, exceptional forms of fossilization in which most of the soft parts have been preserved, and these few examples have given a powerful insight into the detailed anatomy of these extinct arthropods. We shall deal with such details in section 3.2, limiting this introduction to the more apparent characters of trilobite morphology and life habits.

For the description of trilobite morphology and nomenclature we shall refer to the reconstruction of a trilobite in figure 2, representing the dorsal view of the carapace of *Paradoxides gracilis* (Boeck), a beautiful trilobite from the Middle Cambrian of Bohemia. Shown in plate 1 is the photograph of the original specimen, on which the reconstruction has been based. We recognize immediately the bilateral symmetry of the trilobite body, characteristic of typical arthropods. Here the strongly convex axis takes the name of *axial lobe* and the two more flattened adjacent regions are called *pleural lobes.* The pleural lobes are separated from the axial lobe by two *axial furrows.* It is from this longitudinal trilobation (separation into three lobes) that the name "Trilobita" originated and *not* from the transversal subdivision of the body into the three regions of the *cephalon,* the *thorax,* and the *pygidium,* as is often erroneously supposed. The latter subdivision is in fact a character that trilobites have in common with most arthropods. The reader interested in a more comprehensive coverage of the topics which by necessity are mentioned only briefly in this context is referred to the *Treatise on Invertebrate Paleontology* (Moore 1959).

3.1.1. The Cephalon

The head shield or cephalon resulted from the fusion of a number of somites (five or seven) and often carries telltale memory of the original segmentation. It is the most significant and characteristic part of trilobite morphology.

The outline of the cephalon may be semicircular to ogival in its anterior portion, straight or gently curved at the posterior margin, where articulation with the first thoracic segment occurs. The lateral and anterior margin of the cephalon are inflected into the *doublure,* a narrow strip of the exoskeleton which is turned under toward the ventral side. Attached to the anterior part of the doublure is a small shield called *hypostome*, one of the few hard parts to be found on the underside of the trilobite. The angle between the backward-sloping lateral margin of the cephalon and the posterior margin, is called the *genal angle*. This termination can be rounded or, as in our example in plate 1, prolonged into long *genal spines*. The axial lobe extends into the cephalon, where it takes the name of *glabella*. This can be a very convex bulging region, sometimes extending all the way to the anterior margin, or it can terminate earlier, defining a flat *preglabellar field*. The glabella is often sulcated transversally by furrows which may delineate an *anterior lobe* and several pairs of *lateral glabellar lobes*. This particular structure together with the *occipital ring* provides suggestive evidence of the original segmentation of the cephalic region, and is apparent in the trilobite chosen for our reconstruction. The furrows which surround the glabella take the name of the particular area where they are located, as indicated in figure 2. (This is a general rule for furrows occurring elsewhere in the trilobite exoskeleton.) On the sides of the glabella we find the *cheeks*, which are split into two regions, the *free cheeks* and the *fixed cheeks,* by the *facial sutures*. The assembly of the two fixed cheeks together with the glabella constitutes the *cranidium*. About the middle of the interior margin of the free cheeks is a kidney-shaped elevation, the *palpebral lobe*. The visual surface of the eye is usually located between the palpebral lobe and the inner margin of the free cheek. Much will be said about the eyes of trilobites in section 3.3, so we shall not go into details here.

There are several junctions, or *sutures*, between the various parts of the cephalon. The facial sutures mentioned above have been considered meaningful for classification until recently, when other overriding criteria

Fig. 2. Description of trilobite terminology. (a) Dorsal view of the complete exoskeleton. (b) Ventral view of cephalon. The trilobite represented is *Paradoxides gracilis* (Boeck).

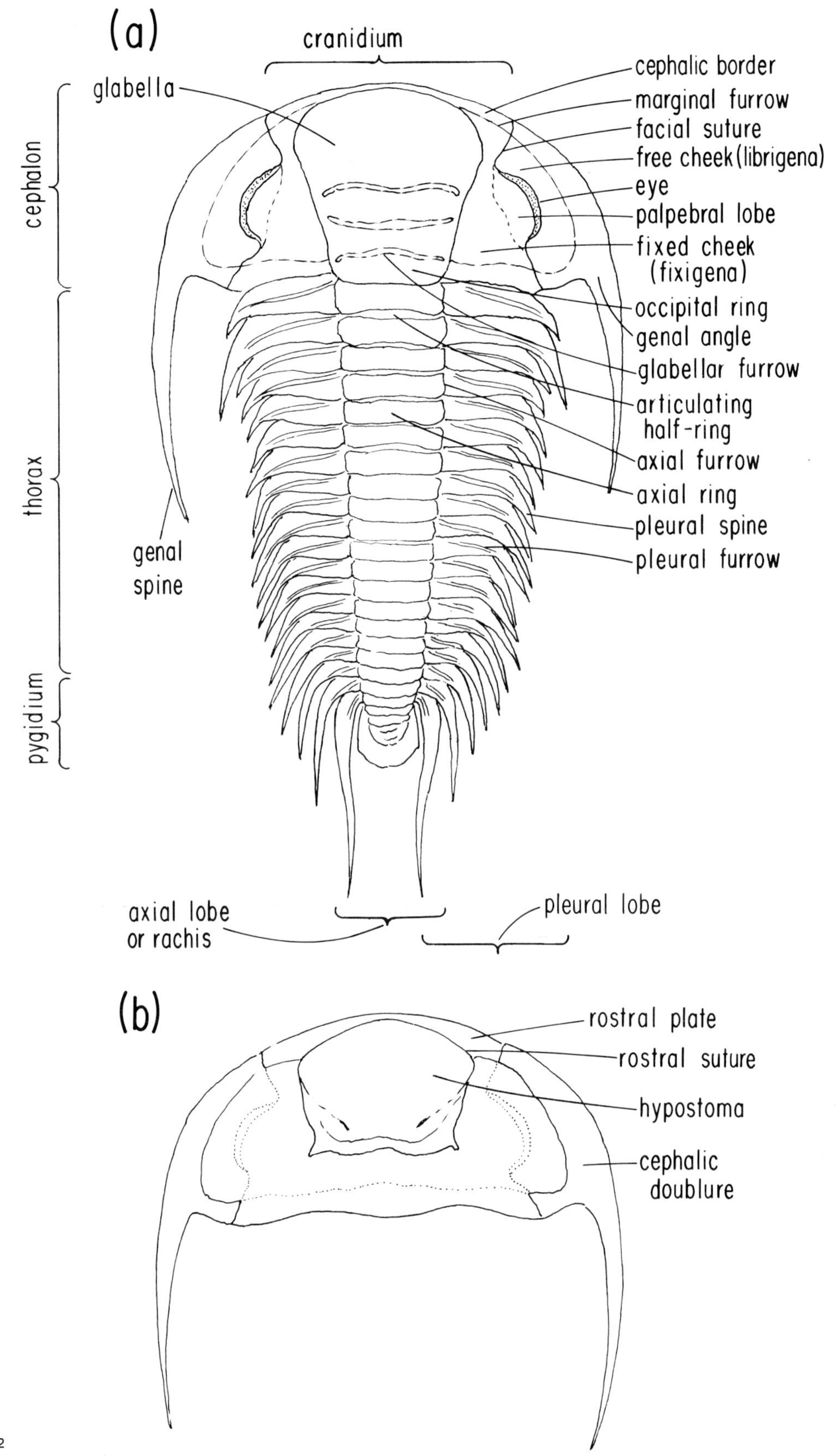

Fig. 2

have been developed. Based on the position of the facial sutures, the trilobites are termed *opisthoparian* if the suture cuts the posterior margin, *proparian* if the sutures run to the lateral margin. In the former case, the genal angle and spine are carried by the free cheeks, in the latter by the cranidium. Figure 2 shows an opisthoparian suture. The free cheeks often separate easily from the cranidium, and this feature may have facilitated molting. A third group exists in which the suture terminates at the genal angle, termed *gonatoparian* (e.g., *Calymene*). As we shall see, the cephalon of trilobites has differentiated into a variety of forms which are characteristic of particular groups and much of trilobite classification is based on cephalic features.

3.1.2. The Thorax

This is the section of the exoskeleton where separate segments are articulated with each other to enable flexibility and enrollment. The number of thoracic segments varies between two and as many as forty-two, most frequently in the range of eight to fifteen. Each segment consists of a center part—the *axial ring*—and two adjacent *pleurae*. These may terminate bluntly or may arch backward into *pleural spines* of varied length. The pleurae are sulcated by a *pleural furrow,* which may have served the function of strengthening the segment. The articulation of one segment with the other occurs through an anterior extension of the axial ring which is inserted beneath the posterior margin of the next segment. The interlocking machanism between segments and the extent of rotation allowed evolved considerably between the earlier trilobite forms and more advanced ones. This function obviously determines the mode and capability of enrollment, which will be discussed in section 3.4.

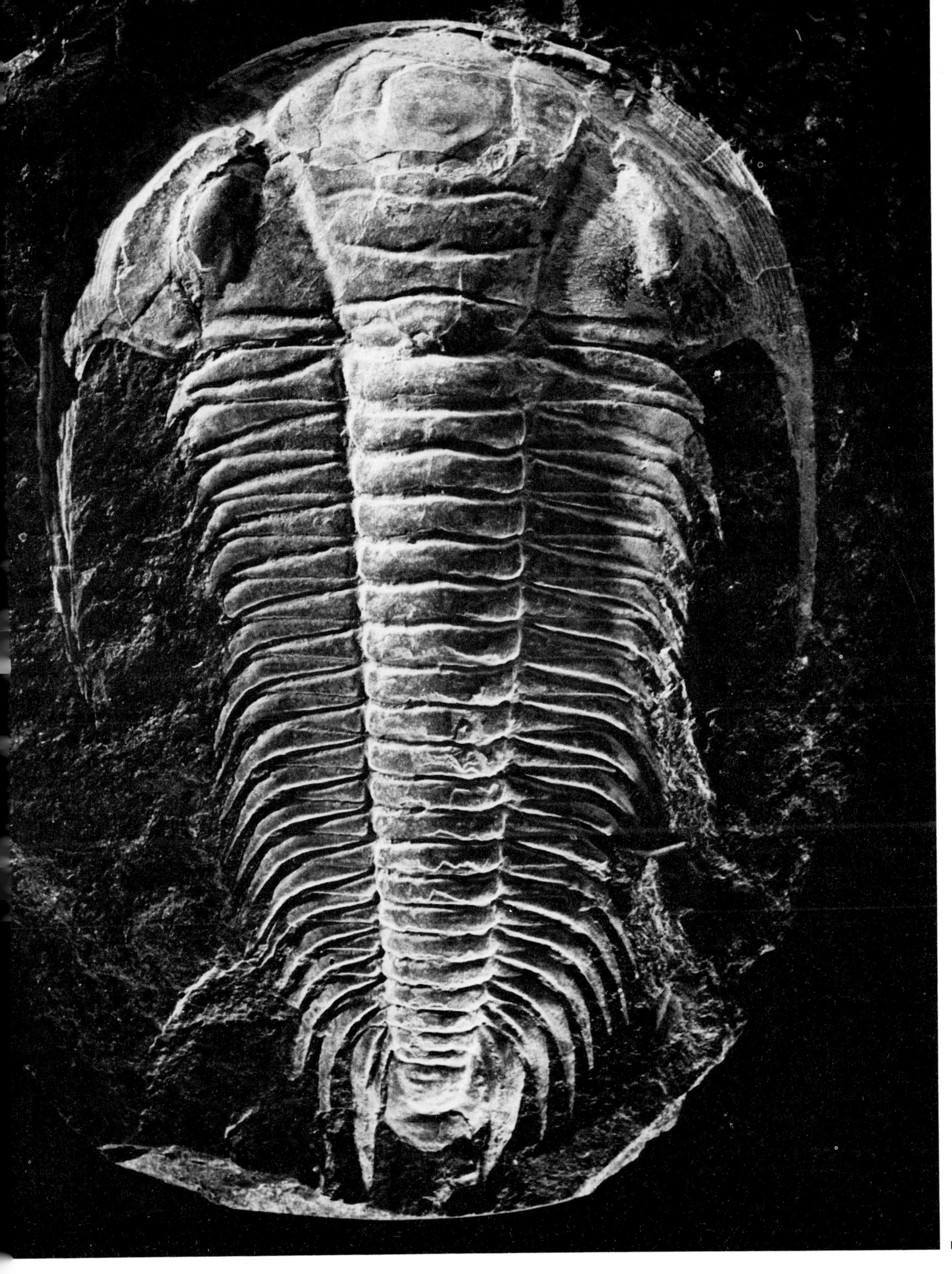

Plate 1. Dorsal view of a complete specimen of *Paradoxides gracilis* (Boeck), a Middle Cambrian trilobite from Jinetz, Bohemia (x2). The reconstruction in figure 2 is based primarily on this specimen, loaned by the Museum of Comparative Zoology, Harvard University.

Pl. 1

3.1.3 The Pygidium

The fusion of a number of somites in the abdominal region gave rise fo the *pygidium*. Extreme variations in size occur for this portion of the carapace, and the extension of the pygidium seems related to the number of thoracic segments. In the example in figure 2 the pygidium is small—the trilobite is called *micropygous*—but the number of thoracic segments is relatively large (twenty for the example shown). Frequently the pygidium reaches size comparable to that of the cephalon (*isopygous* trilobites) as will be seen from many examples in the Atlas. In such cases the number of thoracic segments is usually small. *Macropygous* trilobites are provided with pygidium larger than the cephalon. The axial lobe clearly extends into the pygidial region in a great majority of trilobite species, so that the trilobation is usually preserved. It may extend to the posterior margin or terminate earlier. Evidence of the segmentation from which the pygidium was derived is to be found in the frequently observed ribbing of the axial lobe, resembling the articulated axial rings of the thorax. This causes the axial lobe to appear continuous from the thorax through the pygidial region. There are pleural regions resembling thoracic pleurae, sulcated by furrows. When the latter do not reach the lateral margin, a smooth border results. The marginal border is turned under to form a doublure, much as in the cephalic region.

The shape and ornamentation of the pygidium may reach extravagant extremes. *Marginal spines* are often present, which may be related or not to the pygidial pleurae. Occasionally the pygidium may terminate with a long *axial spine,* as the continuation of the pygidial axis.

3.1.4. Growth and Molting

For a number of trilobite species the various stages of growth (*ontogeny*) from the larval to the adult form are known with great detail from the fossil record. Three major periods of growth are recognized. The *protaspid period* extends from the hatching of the egg to the first appearance on the single-piece dorsal shield of a transverse suture, defining the cephalon and the so-called *transitory pygidium*. During this period a *larval ridge* may be the precursor of the axial lobe. The size of the protaspis is very small, typically 0.3–1.0 millimeters, and the lower limit for any species is clearly the upper limit for the size of the eggs of that species. During this period, the protaspis may develop features, such as marginal spines, which will disappear at a later stage. Observation of the fossil record of growth in the protaspis period has led to speculations concerning the affinities of trilobites with other arthropods, based on the notion that larval development recapitulates ancestral history or phylogeny.

The *meraspid period* is characterized by gradual separation of the cephalon from the transitory pygidium, by means of the progressive appearance of thoracic segments. Meraspid *degrees* correspond to the number of segments which made their appearance. The new segments originate at the thoracic-pygidial boundary. The number of molts required to complete the thorax does not necessarily correspond with the number of segments to be added to the thorax. The overall size of the trilobite increases up to more than ten times that of the protaspis. The meraspid period terminates when the thorax has reached the number of segments which characterize the adult individual.

At this stage the adult form, or the *holaspid period,* has been attained. Here the growth is continuous through many moltings, so that the size of the individual is correlated with its age. Several changes in the relative size of the various parts of the exoskeleton do occur throughout this period. The cephalon usually represents a larger fraction of the carapace in the early growth stages.

The ontogeny of trilobites is beautifully illustrated in an article by H. B. Whittington in the *Treatise of Invertebrate Paleontology* (Moore 1959). We will limit here the illustration of this aspect of trilobite life to the presentation of a pictorial summary in plate 2. This picture contains evidence of some kind of trilobite nursery, containing examples of all stages of growth. The trilobite depicted is *Elrathia kingii* (Meek) from the famous Wheeler formation of the Middle Cambrian of Utah. Although detail may be lost in the smallest specimens, this view gives a feeling for the range of sizes involved in trilobite growth. Present are several holaspid carapaces of various length, with their characteristic number of thirteen thoracic segments; at least one meraspid carapace with eight segments; and at least one protaspid shield about one millimeter long, showing the larval ridge quite distinctly.

Molting is clearly an integral part of the growth process in trilobites. The most abundant fossil remains of trilobites are the disarticulated exuviae, which, on account of their relatively large surface and light weight, could be easily transported and concentrated by wave motions and currents. Plate 3 shows a slab of Ordovician shale

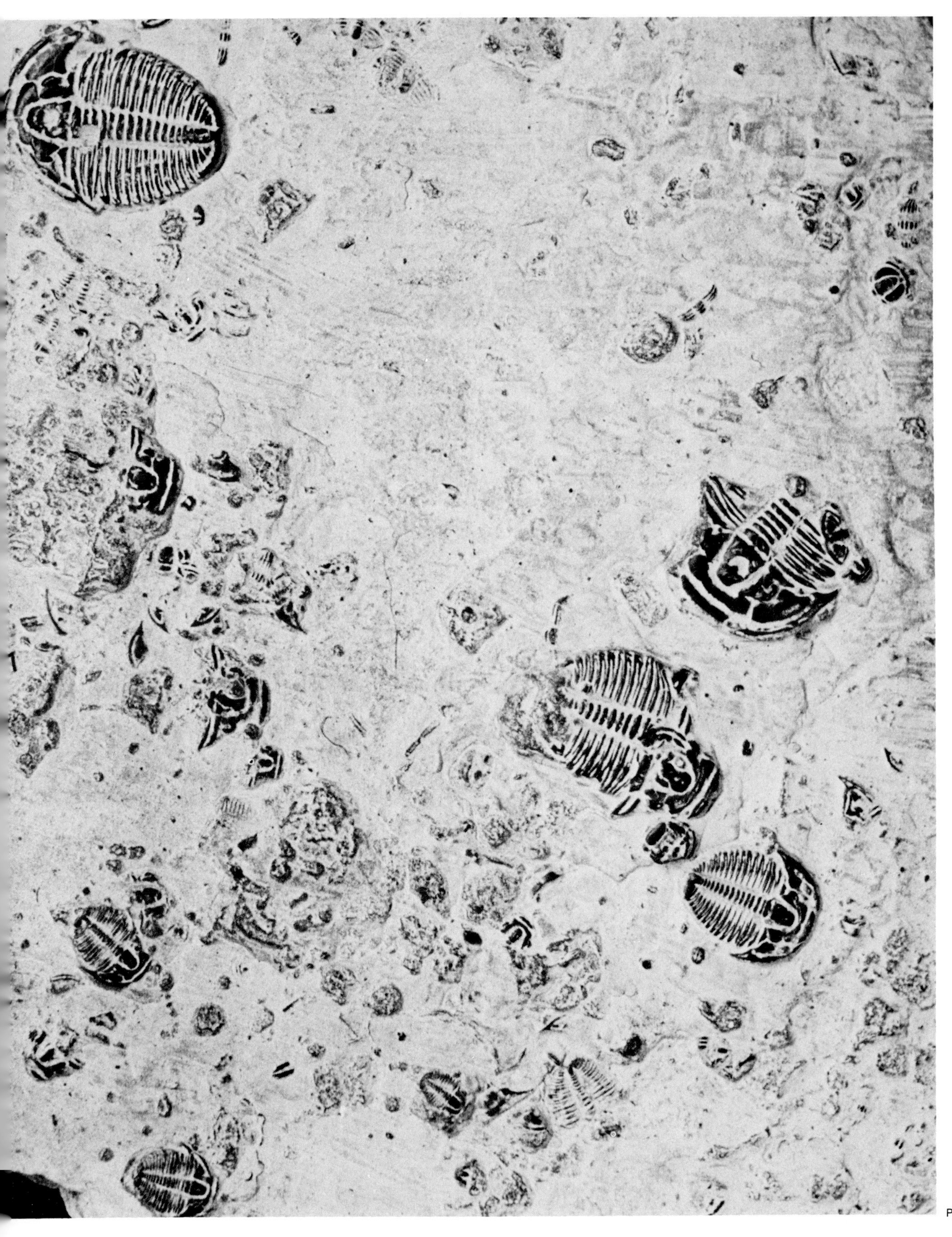

Plate 2. Various stages of growth of the trilobite *Elrathia kingii* (Meek) from the Wheeler formation, Middle Cambrian, Utah (x3). (RLS coll.) The larger trilobites represent the holaspid stage, a meraspid carapace with eight segments is the center lowermost trilobite, a protaspid shield is located just below the lowermost complete trilobite on the right hand side.

Pl. 2

Plate 3. Slab coated with disarticulated exuviae of the trilobite *Ogygites canadensis* (Chapman) from the Collingswood formation, Ordovician, Collingswood, Ontario (xl.2). (RLS coll.)

Pl. 3

from the Collingswood formation, Ontario, which is densely covered with a multi-layer of carapace fragments of *Ogygites canadensis* (Chapman). Such accumulation is most likely due to selective concentration, as is often seen occurring with small bivalve shells on the sea shore. In such occurrences complete carapaces are very seldom found. Since trilobites molted many times during their growth, it follows that each individual left a multiple fossil record.

In situations where the molting process occurred in a relatively undisturbed environment, exuviae may be found in the approximate posture in which they have been adandoned by the trilobite. There are at least two characteristic modes in which the exoskeleton was shed by the growing trilobite. In one of them, the *phacopid mode*, the cephalon separated from the thorax between the occipital ring and the first thoracic segment. The animal would crawl out of the old exoskeleton through such an opening, and in so doing would force the old cephalic shield to flip upside down. Plate 4 shows one of many examples, collected by the author, of exuviae shed by *Phacops rana milleri* Stewart, as found in the Devonian Silica shale at Sylvania, Ohio. The cephalon is intact, and usually is found in the immediate vicinity of the thorax-pygidium assembly. The latter occurs most often in the enrolled condition, showing that the integument must have contracted like a spring, after the former inhabitant crawled out. As mentioned previously about arthropoda, the newly molted animal is provided with a soft exoskeleton, which would later harden through mineralization. Plate 5 shows a "soft-shelled trilobite," also from the Silica shale. We are dealing here once again with *Phacops rana milleri* Stewart. This soft shell condition is revealed by the fact that the carapace is considerably thinner than in the average specimen. This gives a translucent quality to the shield, particularly apparent in the pleural regions. The axial lobe seems to have thickened somewhat more and appears darker. By accident or not, this trilobite overlaps the exuviae of another trilobite (or its own?).

The above description seems to ignore the fact that we are dealing here with a fossil and not with a living animal. However, here the fossilized exoskeleton is a faithful calcite replica of the original one. Furthermore, the color differentiation which affects selected areas of the carapace in the example in plate 5 may in fact reflect incomplete mineralization of the exoskeleton as it existed at the time of burial.

Another pictorial view of phacopid molting is contained in plate 6. Here we deal with exuviae of *Dalmanites verrucosus* (Hall) from the Silurian Waldron shale formation of Waldron, Indiana. In this example all trilobite parts are seen from the ventral side, and are still partly imbedded within the shale matrix. The hypostome appears displaced from its axial position, indicating breakage of the hypostomal suture during molting. In the exuviae shown in plate 5 the hypostome is in fact missing altogether.

Another frequent molting procedure is the *olenid mode*, in which the free cheeks separate from the cranidium at the facial sutures, enabling the molting trilobite to make its way out of the opening thus created. In this case the facial sutures are functional, representing natural fracture lines.

In spite of the predominance of exuviae in the fossil record of trilobites, an animal would occasionally be buried intact, as the majority of the pictures in the Atlas section will illustrate.

3.2. Appendages and Internal Anatomy

Although the more commonly preserved part of the trilobite body is the hard, mineralized carapace, there are a few fossil deposits which have yielded completely preserved bodies, from which, by proper techniques, even the soft parts of the internal structure can be recognized. Famous formations of this kind are the Middle Cambrian Burgess shale of British Columbia, where the ventral appendages appear as flattened, extremely detailed impressions; the Middle Ordovician Utica shale of New York; and the Lower Devonian Hunsrück shale of Germany. In the fossils from the latter two deposits, the trilobite soft parts are replaced by fine crystals of iron pyrite, and this fortunate feature lends itself to detection by visual observation and, more effectively, by radiography. Trilobites with preserved appendages have occasionally also been found in very fine-grained limestones. From the above occurrences, the ventral appendages of not more than twenty species of trilobites have been described in several famous studies by Walcott, Raymond, Størmer and others. These studies are comprehensively summarized in the *Treatise* (Moore 1959).

In recent years improved x-ray techniques have yielded much new information on the subject, in particular from studies by Stürmer and Bergström (1973) on the trilobites from the Hunsrück shale, and by J. Cisne (1973) on the trilobites from the Utica shale. I had the fortune of working in close collaboration with Dr. Cisne when he

Plate 4. Exuviae of *Phacops rana milleri* Stewart, from the Silica shale formation, Devonian, Sylvania, Ohio (x5). (RLS coll.)

Pl. 4

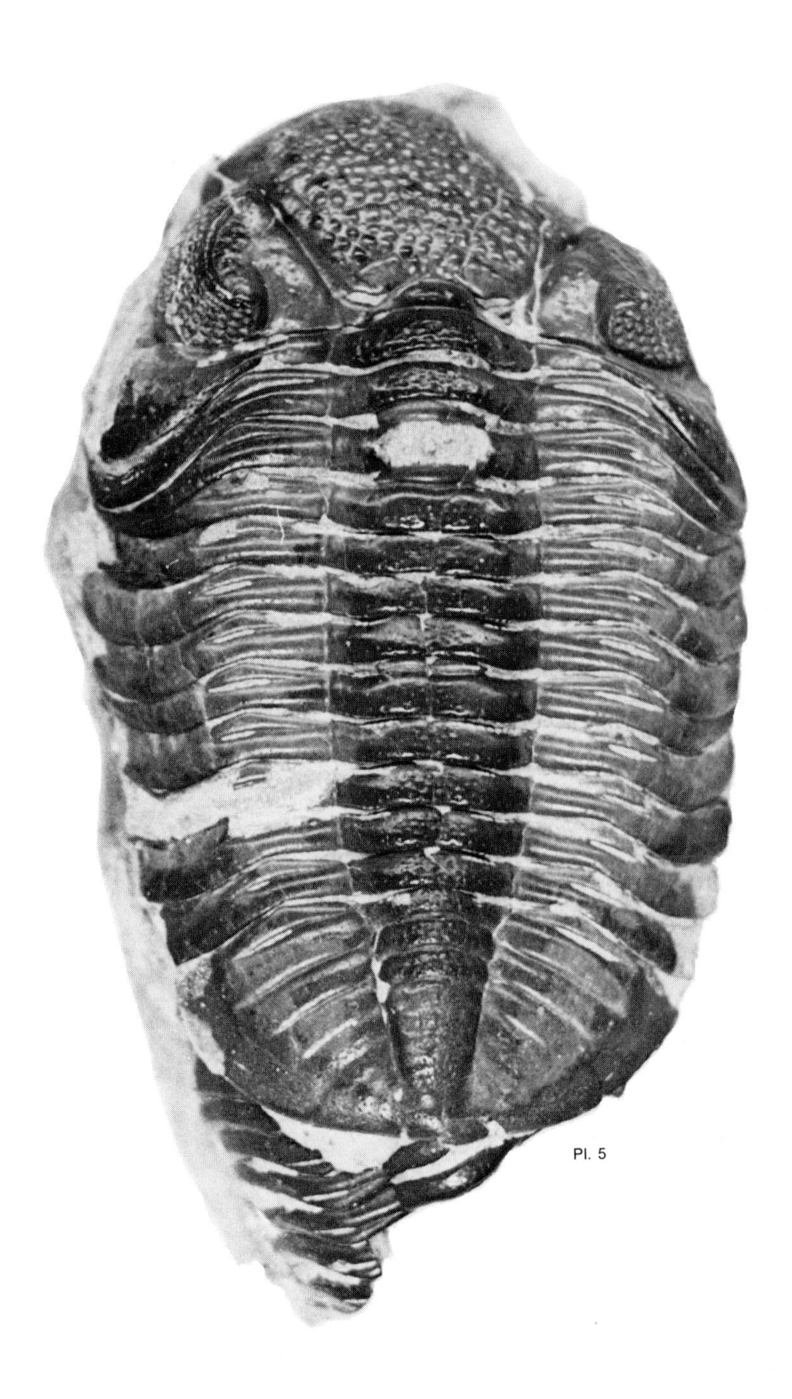

Pl. 5

Plate 5. An example of "soft-shelled trilobite" (x3). Due to the unusually thin and translucent carapace, this example of *Phacops rana milleri* Stewart, is interpreted as representing a newly molted animal. Silica shale, Devonian, Sylvania, Ohio. (RLS coll.)

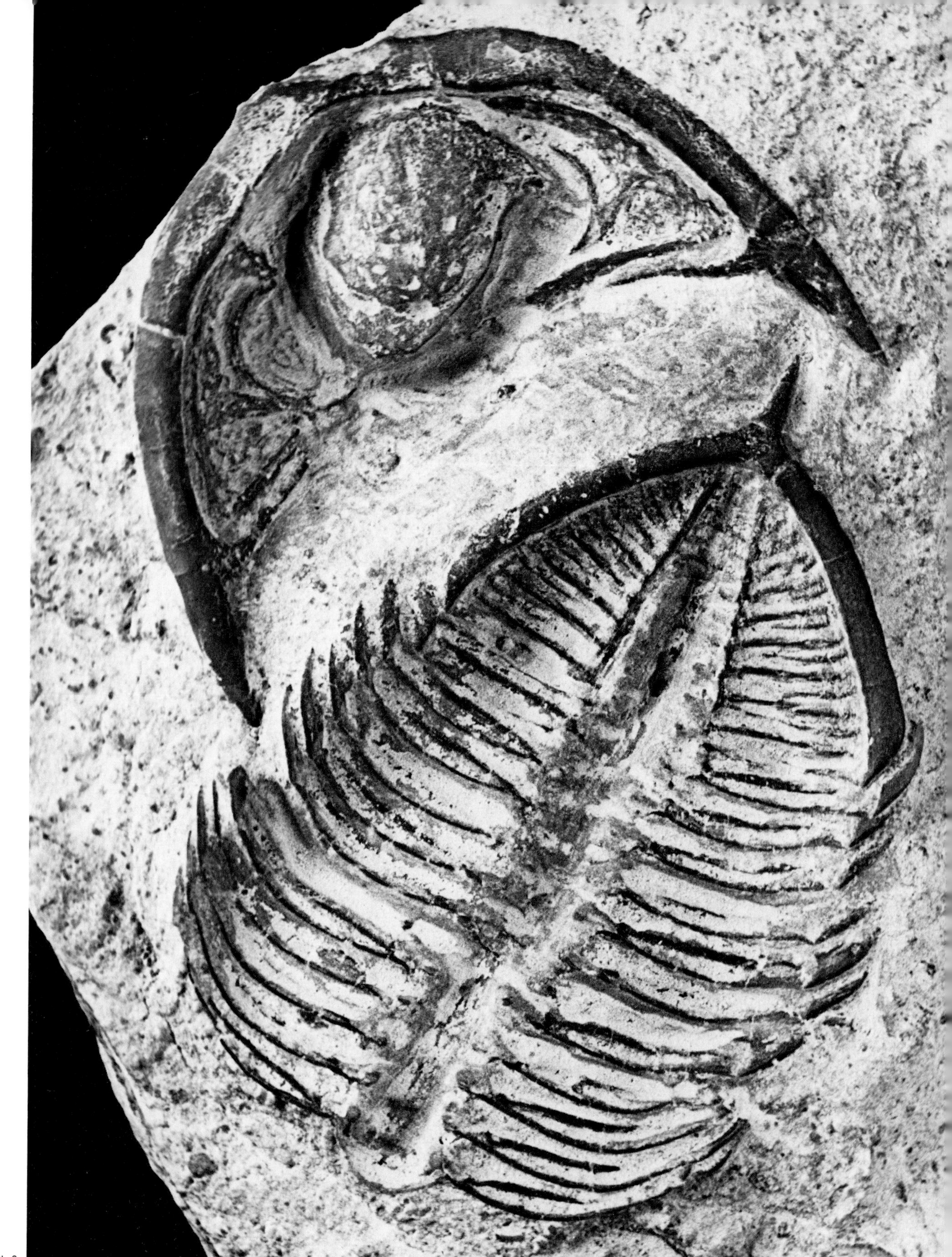

Plate 6. Exuviae of *Dalmanites verrucosus* (Hall), seen from the ventral side (x4.2). Silurian, Waldron shale, Waldron, Indiana. (RLS coll.) Note the displaced hypostome.

Pl. 6

was preparing his Ph.D. dissertation on the anatomy of *Triarthrus eatoni* (Hall), the most abundant of the trilobite species in the Utica shale. The material at the disposal of Dr. Cisne was the exceptional collection of specimens assembled and prepared by C. E. Beecher toward the end of the last century and now belonging to the American Museum of Natural History, Field Museum of Natural History, Harvard Museum of Comparative Zoology, and the Yale Peabody Museum. The trilobites originate from a thin layer, about 10 mm thick, in the black Utica shale. Beecher quarried this layer extensively, collecting some 700 specimens, and exposed the appendages of some 60 specimens by gently rubbing away the matrix with pencil erasers. Several trilobites were further prepared by Dr. Cisne to make them suitable for soft x-ray examination, which was carried out by taking stereoscopic radiographs. In addition to furthering a better understanding of the already known appendages, Dr. Cisne's study revealed the presence of previously unsuspected details of the internal anatomy of the trilobites. As it turns out, the entire digestive tract of these specimens was often preserved, as well as muscle fibers and other surprising features of the trilobite's body. Several of Dr. Cisne's original radiographs were kindly loaned to the author for printing and are presented in this book (see plates 8, 10, 11, 63a, 64). Furthermore, several specimens were selected and loaned to enable the author to experiment with the optical techniques most suitable to enhance the visual contrast of the exposed appendages. Some of the photographs in this section and later in the Atlas constitute a unique record of Beecher's trilobites.

Figure 3 presents a reconstruction of the dorsal and ventral side of *Triarthrus eatoni* (Hall), based on the latest observations by J. Cisne. It differs from previous reconstructions in several details, the most prominent being the number of cephalic appendages (three instead of four pairs), and the presence of an abdominal extension carrying the anal tract. The basic feature of the ventral anatomy is the presence of a pair of *biramous appendages* carried by each of the thoracic segments. The first pair is modified into two segmented *antennae,* which serve an obvious sensory function. The cephalic region also contains three pairs of slightly modified appendages, and the pygidium five pairs, while about eight pairs are carried by the post-pygidial abdomen. In total, approximately thirty-one pairs of limbs can be recognized. The basic structure of the biramous appendages is shown in Figure 4. It is in many respects similar to that found in certain crustacea. The base portion of the limb, the *coxa*, carries two differentiated branches: a feather-like *exite* and a *telepod* composed of seven articulated limb segments. Short bristles or *setae* appear at various locations. The featherlike structure of the exite has been interpreted as representing gill-blades, which performed the respiratory function. This interpretation, however, is still controversial. Nevertheless, there is no doubt that telepods were constructed for crawling. The tracks left by the motion of the telepods on the soft sea floor are preserved occasionally as fossils trails, called *Cruziana.* The exites could have helped in swimming, their shape and arrangement suggesting a "venetian blind" oarlike stroke.

Evidently trilobites were not provided with claws or mandibles. They were filter-feeders, as are many crustaceans today. Their food, in the x-ray plates of J. Cisne, can occasionally be observed as clouds of finely particulate material squeezed out of the gut canal following burial compression. The mouth is a small opening, posteriorly oriented, just at the tip of the hypostome. It is the terminal feature of the long enclosure defined by the limb coxae. This suggests a so-called trunk-limb feeding mechanism. It involves the creation of a feeding current by a rhythmical motion of the limbs, in this case making the coxae convey the food particles to the mouth through the ventral food groove. The bases of the head limbs were somewhat differentiated from the thoracic limbs, indicating some kind of weak masticatory function. Locomotion, feeding, and mastication were probably part of a mechanically related sequence of events produced by the movement of the limbs. Once ingested, the minute food particles would pass through an esophagus into the stomach, located beneath the frontal glabellar lobe, and then into the intestine, a long tube running through the axial region and terminating in the anal duct.

Once considered as part of the digestive system and called *genal caeca* is a network of ramifications radiating away from the axial region, into the genal area of the cephalon. This anatomical feature gives rise to the *prosopon* (functional ornamentation) observed on the dorsal cephalic surface of many trilobites as a fine mesh of radiating and ramifying ridges. It is now believed to represent part of the vascular or circulatory system (Bergström 1973) and will be repeatedly noticeable in the Atlas photographs.

Another marvel of Beecher's trilobites is the preservation of the muscle structure. As observed by Dr. Cisne, this consisted of a very efficiently engineered network of longitudinal, dorsoventral, horizontal, and limb muscles, which would ensure articulation,

Fig. 3. A modern reconstruction of *Triarthrus eatoni* (Hall), completed from partial drawings by Cisne (1973). Details of the ventral view are occasionally omitted to show underlying structure.

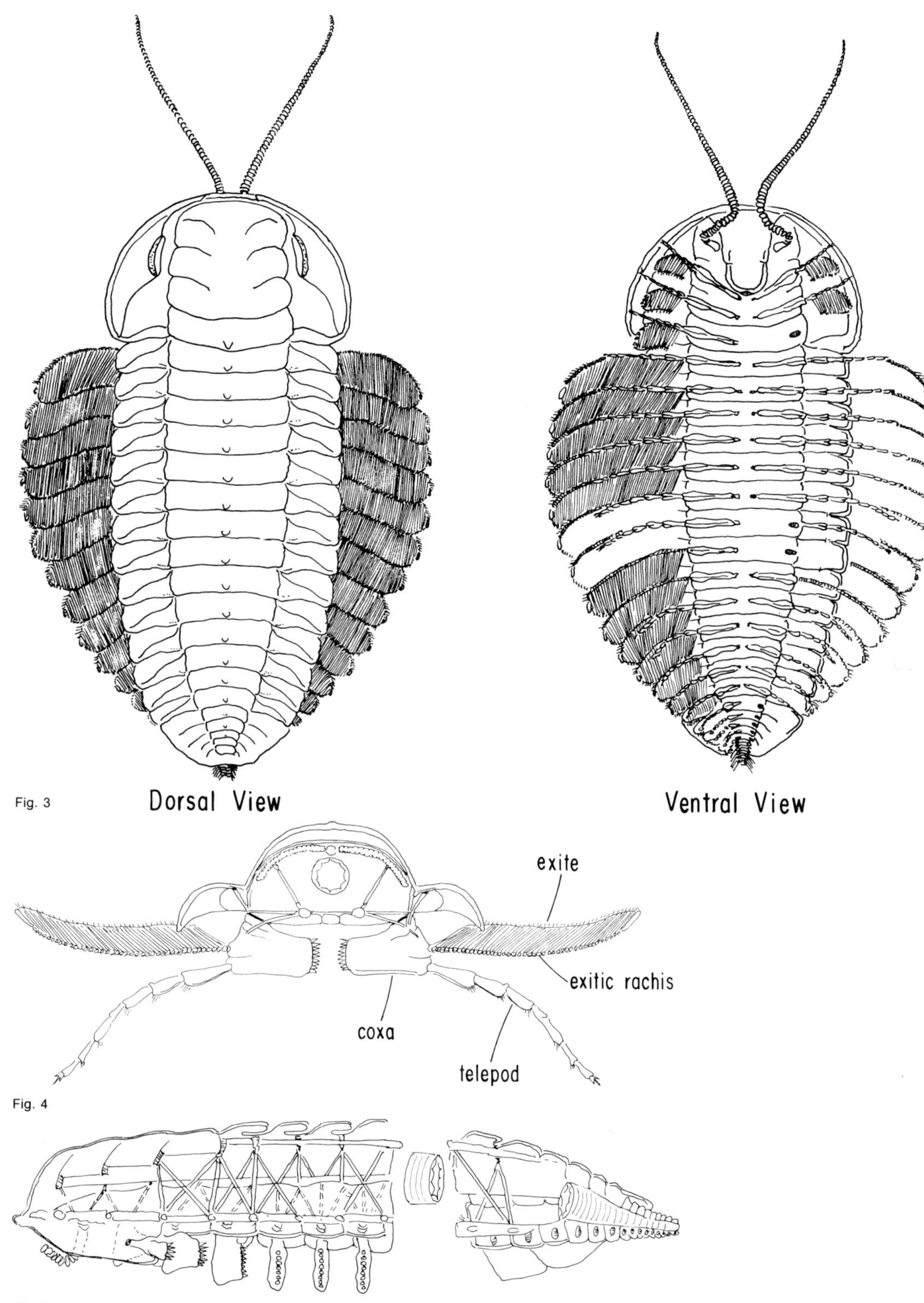

Fig. 4. Structure of the biramous appendages and cross sectional view of the thoracic region in *Triarthrus eatoni* (Hall). Combined assembly from separate reconstructions by Cisne (1973).

Fig. 5. Longitudinal section through parts of the body of *Triarthrus eatoni* (Hall). This drawing, adapted from reconstructions by Cisne (1973), shows details of the muscle structure and of the intestinal duct.

limb movement, and enrollment, and which more generally were designed to hold together the exoskeleton. Figure 4 shows only a few of such muscles in a transverse cross section of the thoracic region. Figure 5 shows portions of a longitudinal cross section, indicating, on the one hand, the complexity of the trilobite design but, on the other, its rationality.

Many details of other organic functions have been revealed in Dr. Cisne's study. We will limit our description here, however, to a presentation of some of the material which led to so much insight into the anatomy of *Triarthrus.* In plate 7 we see the ventral side of a specimen of *Triarthrus,* as originally prepared by C. E. Beecher. The photograph has been taken with the specimen totally immersed in xylene (see Appendix B). The optical contact established by this liquid of high refractive index eliminates surface reflections and enables the achievement of maximum optical resolution. The hypostome, the antennae, and most of the appendages can be seen. The pyritized organic material appears white against the black matrix.

Plates 8 and 9 represent another specimen of the same trilobite as seen in, respectively, a radiograph by Dr. Cisne and a photograph by the author. It must be remembered that the radiographs were taken in stereo pairs. The stereoscopic observations of such pairs enables the distinguishing of features occurring throughout the depth of the specimen. Such features overlap in a single projection and make their interpretation more difficult. The appendages are clearly visible however, and can be seen even when underlying the dorsal shield. Although the radiograph has very high resolution, the featherlike structures of the exites is not as apparent as in plate 9, which was obtained by immersing the specimen in

Plate 7. Ventral view of a specimen of *Triarthrus eatoni* (Hall), whose pyritized appendages have been exposed after painstaking preparation by C. E. Beecher (x5.2). From "Beecher's trilobite bed," Frankfort shale, U. Ordovician, Rome, N.Y. (YPM 219, loaned through courtesy of Dr. J. Cisne.) Photograph obtained while the specimen was totally immersed in xylene.

Pl. 7

Plate 8. Radiograph of an adult specimen of *Triarthrus eatoni* (Hall) (x4.2). (YPM 228, from a stereo pair of radiographs by J. Cisne.) The biramous appendages can be seen protruding from and underlying the dorsal shield. Details of other soft parts, replaced by fine granules of iron pyrite, yield visible contrast.

Pl. 8

Plate 9. The same specimen as in plate 8, this time immersed in xylene. Fine surface details of the structure of the appendages are revealed by this technique.

Pl. 9

Plate 10. Another x-ray view of a completely preserved specimen of *Triarthrus eatoni*, as in the preceding plate (x8.2). (YPM 28253, radiograph by J. Cisne.) These pyritized trilobites represent one of the most striking fossil records of extinct life ever recaptured.

Pl. 10

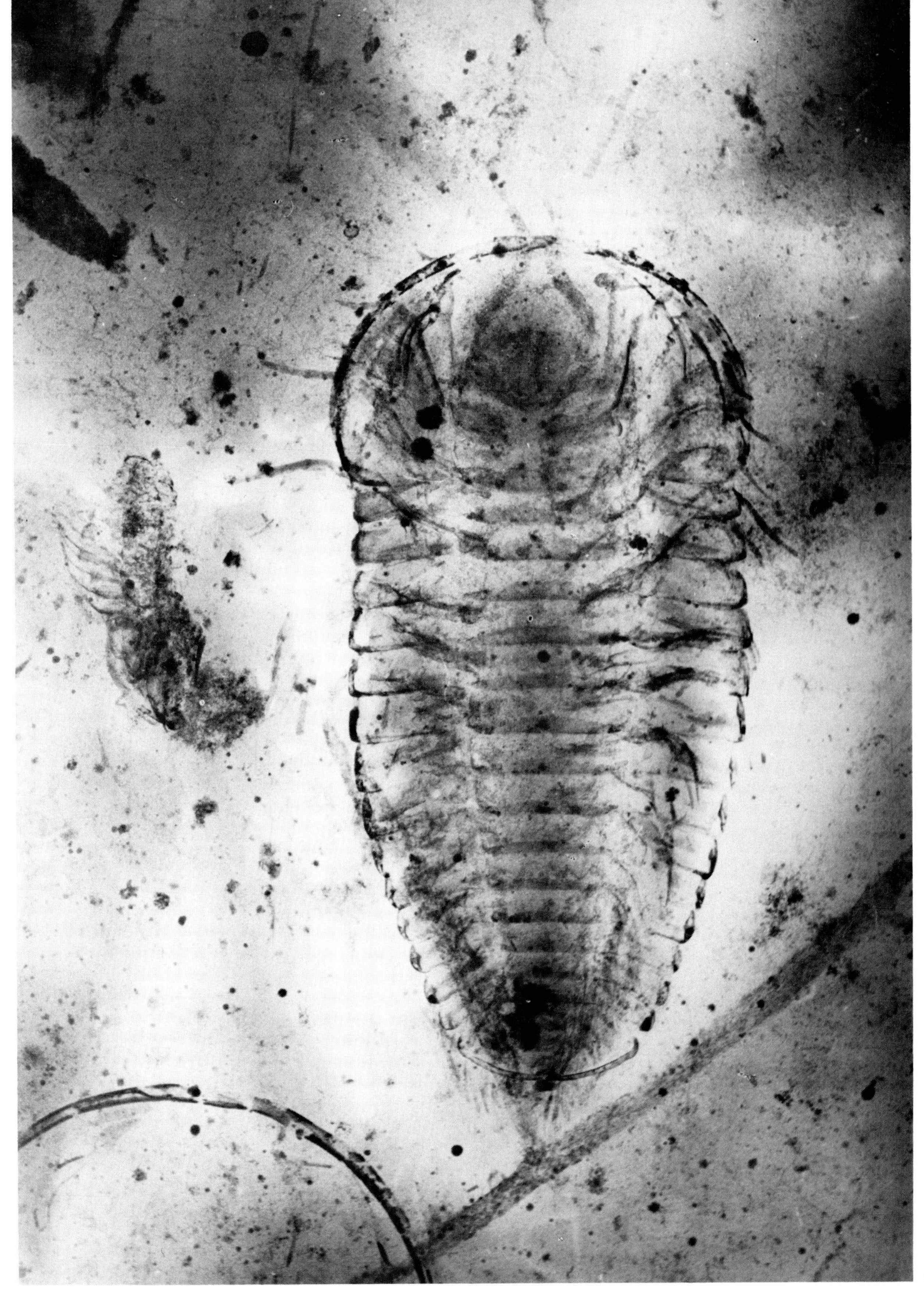

Plate 11. The world of *Triarthrus* seen once more through a radiograph by J. Cisne (x8.2). (MCZ 7190/14 A–C)

Pl. 11

Plate 12. Dorsal and ventral view of reconstructions of *Olenoides serratus* (Rominger), a Middle Cambrian trilobite. The models were prepared at the Paleontological Museum, University of Oslo. Print from a color slide.

Pl. 12

a bath of xylene, as for plate 7. The photographic technique employed here is showing its advantages. The exitic filaments are now clearly discernible together with the segmented structure of the exitic rachis, which carries them.

Plates 10 and 11 are radiographs of two other specimens. Here the digestive tract is visible, in particular in the vicinity of the anal termination. Many of the anatomical details mentioned previously can be made out from these examples, in particular with the aid of a magnifier. It should be remembered however that no individual specimen will show all the features exhibited in the reconstructions illustrated in Figure 3. It is only through the examination of a large number of specimens that the overall structure can be visualized.

Other examples of *Triarthrus* will be shown in the Atlas section, where the taxonomic information concerning this trilobite will also be given. Much more is contained in Dr. Cisne's dissertation, in particular very important observations relating trilobites to the other groups of arthropods. Further discussion, however, exceeds the scope of this presentation.

In order to conclude this section on a somewhat less technical note, plate 12 shows what trilobites actually may have looked like. The two creatures shown, one in the normal crawling posture, the other helplessly overturned, were photographed by the author on the patio of a beautiful resort in the mountains overlooking Oslo, where the International Conference on Trilobites was held in July 1973. We are dealing here with reconstructions of *Olenoides serratus* (Rominger), prepared at the Paleontological Museum, University of Oslo, and displayed for the enjoyment of the convening paleontologists. In spite of their appearance in this plate, trilobites were not light-emitting animals. The picture is simply a print from a color slide and is therefore what is usually regarded as a negative. These trilobites possessed another pair of sensory appendages, posteriorly located and called *cerci*. They seemed to enjoy the midsummer Norwegian sun.

3.3. The Eyes of Trilobites

Not only were the trilobites first in developing highly organized visual organs, but some of the recently discovered properties of trilobite's eye lenses represent an all-time feat of function optimization. We are confronted here with a very successful scheme of eye structure: the composite or compound eye, made of arrays of separate optical elements, the *ommatidia,* pointing in slightly diverging directions and each performing an identical function. A network of photoreceptors and neurons translates the optical stimuli into an image perception. Evidence of the success of such a scheme is widespread experience, since the eyes of insects and crustacea, in fact of most arthropods, still follow a design closely related to that developed by trilobites.

3.3.1. The Compound Eye

In modern arthropods the structural unit (ommatidium) is made of a sequence of functional subunits (see fig. 6). Facing the outside world is the dioptric apparatus consisting of a *corneal lens* in optical contact with a *crystalline cone.* Tiny images from a narrow field of view appear at the tips of the cones. Proceeding toward the interior of the eye beyond the cone there are two types of structures. In the so-called *apposition eyes* (mostly found in diurnal insects, crustacea, etc.), the photoreceptor, or *rhabdom*, is long and attached directly to the tip of the crystalline cone. On the other hand, in the *superposition* eye (found in nocturnal forms like moths, fireflies, etc.) a crystalline fiber, like a light guide, intervenes between the tip of the cone and a shorter rhabdom. A dark pigment may fill the space between cones in the apposition eyes, while in the superposition eye the pigment layer can migrate to surround either the cones or the crystalline fibers, when the eye is dark- or light-adapted respectively. Furthermore, the superposition eyes are generally constructed with very regular radial symmetry. This feature and the pigment migration in the dark-adapted eyes have been interpreted as enabling occurrence of collective phenomena such as superposition of images or diffraction patterns due to adjacent ommatidia. The interpretation of the visual process in the compound eyes has, however, been the subject of great controversy since the pioneering work of Müller (1826) who introduced the "mosaic" theory, according to which the compound eye is regarded as an assemblage of directional units, each yielding a point element in a mosaiclike reconstruction. Exner (1891) further developed the mosaic theory in his classical study of faceted eyes and introduced the distinction between apposition and superposition eyes. However, most of Exner's models of image formation by the dioptric system of the ommatidia turned out to be wrong. For example, Exner assumed that the crystalline cones in *Limulus* would form images at their tips due to a radial variation of the refractive index (cylinder lens). This is now known not to be true; in fact the same images can be obtained from a homogeneous scaled-up replica of the *Limulus* cone made of lucite when it is immersed in water (Levi-Setti, Park, and

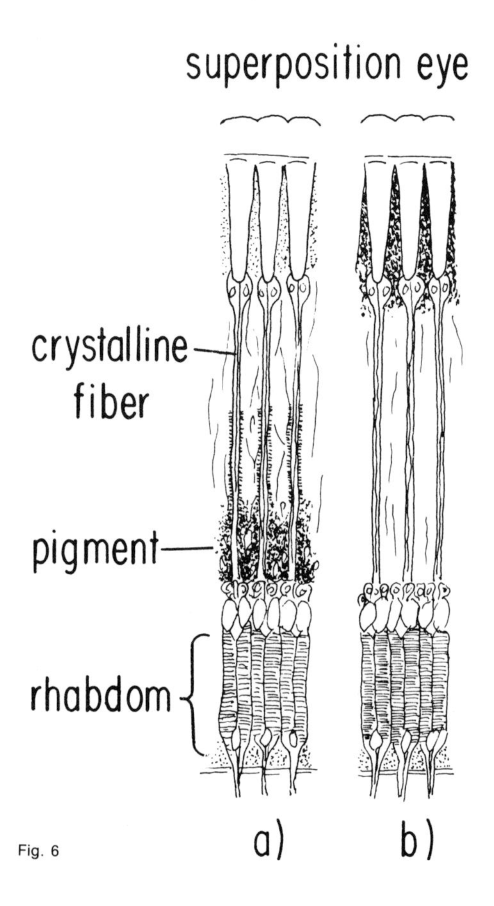

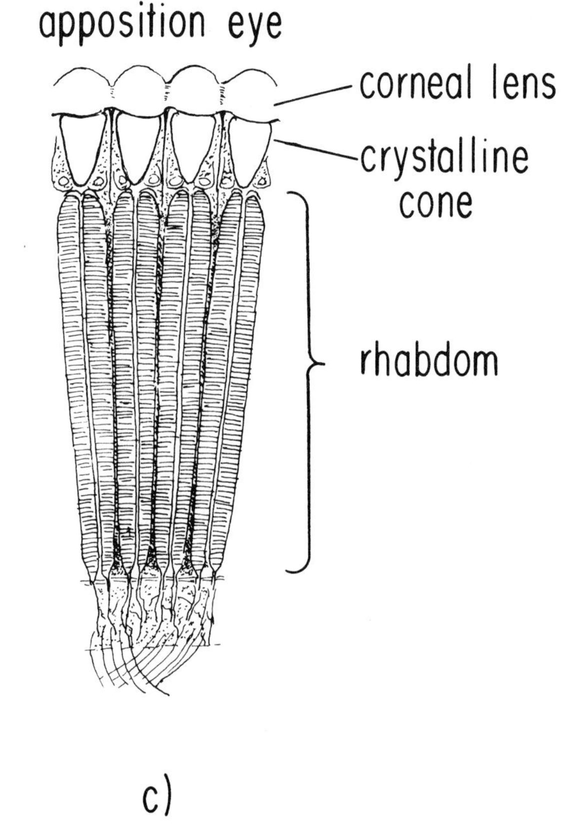

Fig. 6

Fig. 6. Structure of the ommatidia in the compound eye of modern arthropods. Parts (a) and (b) are schematic sections through the ommatidia of the firefly. Pigment migration distinguishes the light-adapted condition (a) from the dark-adapted condition (b) (adapted from Horridge 1969). Part (c) represents the structure of the ommatidia of the ant (adapted from Snodgrass 1952). The two types of eye are distinguished by the presence of short and long rhabdoms respectively.

Winston 1973).[1] Furthermore, Exner ignored the role of the crystalline tracts between cone and rhabdom, present in the so-called superposition eyes. What seems to emerge from modern research is that the term *superposition* is a misnomer (Horridge 1969), since no superposition of sharp images is ever observed in the plane of the photoreceptors when the crystalline tracts are in place. After more than eighty years of research, the only significant distinction found in the types of arthropod eyes is that there are eyes *with* or *without crystalline threads,* and, correspondingly, with short or long rhabdoms. What on the other hand has become increasingly apparent is the role of the neurophysiological apparatus in modifying the type of response which could be inferred from purely optical considerations. The lateral inhibitory interaction can alter the effects of image overlap between neighboring ommatidia and sharpen contrast. (For a summary see, for example, Hartline 1969.) A functional distinction between the two types of eyes discussed above is still obscure but may not be as fundamental as originally thought by Exner.

3.3.2 The Compound Eye in Trilobites

Although we have no knowledge of the internal structure of the eyes of trilobites, the fossil record yields astounding evidence of the dioptric apparatus, the outermost region of the ommatidia. This is due once again to the fact that only this portion was sufficiently mineralized to remain preserved

[1]The shape of each ommatidium of *Limulus* has been found to be optimized for the maximum collection of light incident within a field of view of aperture angle $\pm 19°$ from the axis. The relevant phenomenon taking place in such a device is total internal reflection at the interface between the corneal medium, or refractive index $n = 1.53$, and the fluid external to the cone, of refractive index $n = 1.35$.

Fig. 7. Cross-sectional and frontal views of the visual surface in several holochroal eyes of trilobites. (a) *Sphaerophthalmus alatus* (Boeck), U. Cambrian, Sweden (x80); (b) *Cyrtometopus clavifrons* (Dalman), Ordovician, Sweden (x50); (c) *Illaenus chiron* Holm, Ordovician, Sweden, (x50). Adapted from Lindström, 1901.

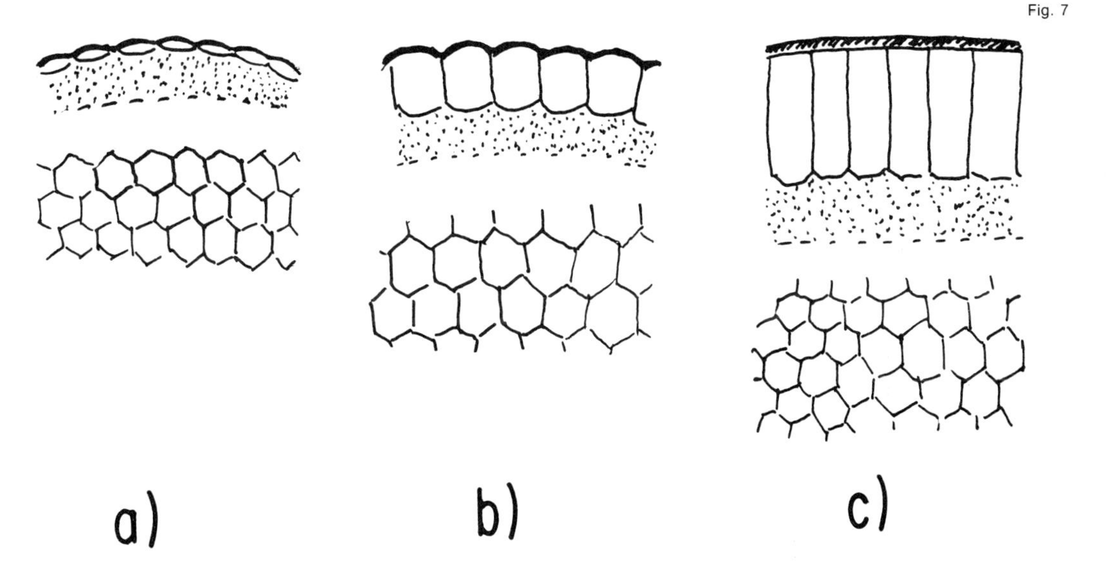

in the fossilization process. In favorable circumstances the eye lenses survived intact their long burial: they were already made of calcite crystals in the living animals! Not all trilobites possessed eyes—some, in fact, were blind; but others had enormous eyes which would take up most of the cephalic surface. Often the eyes were shaped in turretlike fashion, and their combined visual field could cover the animal's entire surroundings. Trilobite eyes, as we shall see, are revealing indicators of the habits of their carriers. Lacking the complete structure of the ommatidia, we cannot, of course, draw an exact analogy between the compound eyes of trilobites and those of modern arthropods. However, the external appearance and the ommatidial arrangement are often suggestive of a very close correspondence. If provided with the present knowledge, Exner would probably recognize apposition and superposition eyes in trilobites. Lindström (1901) instead recognized two different kinds of trilobites' eyes: a truly compound eye and an aggregate eye. The compound, or *holochroal*, eye is characterized by a close packing of the ommatidia, the entire visual surface being covered by a continuous pellucid membrane, the *cornea*. The ommatidia vary in shape from thin biconvex lenses to elongated hexagonal prisms. A different structure is presented by the aggregate, or *schizochroal*, eye. Here the lenses are separately encased and positioned by a cylindrical mounting, the *sclera,* and each lens is covered by its own cornea.

The holochroal eye with thin biconvex lenses is thought to be the ancestral form (Clarkson 1973a), already well developed in the late Cambrian, while the holochroal eyes with thick prismatic lenses and the schizochroal eyes appear in post-Cambrian times. Any close similarities between trilobites' and modern arthropods' eyes is more likely to exist in the ancestral form of holochroal eye, where the two branches of arthropod evolution were phylogenetically closer. As we shall see, there are in fact forms in the holochroal eye which suggest a superposition-type organization. On the other hand, there is no reason to suspect that the apposition-type was not already present in trilobites. The schizochroal eye is in fact externally organized very much in the same way as the latter. As we shall see, however, the Phacopida, the principal possessors of the schizochroal eyes, have evolved a lens structure which is not known to exist in modern arthropods; this is understandable, since it developed when trilobites were already a well-separated stock.

Most of the photographs in this section originate from negatives kindly loaned by Dr. E. N. K. Clarkson of the Grant Institute of Geology, University of Edinburgh, Scotland, who has in recent years carried out comprehensive studies of the visual apparatus of trilobites. The author is presently collaborating with Dr. Clarkson on the study of the optical functions of several types of ommatidial structures, and some of the discoveries emerging from this work will be described here.

3.3.3. Holochroal Eyes

A schematic view of a range of ommatidial structures in this type of eye is shown in figure 7, adapted from Lindström (1901). In addition to the basic hexagonal prism design, other lens contours were present, including square prisms. The optical behavior of these elements, while rather obvious in the simple biconvex lenses of *Sphaerophthalmus*, is not so intuitively clear in the case of the long prisms of *Asaphus* and *Illaenus*, in particular since such prisms were made of single calcite crystals (Clarkson 1973b). To offset the strong birefringence of calcite, the crystals were oriented so that the optic axis always pointed in a direction normal to the visual surface. Only along this axis does calcite behave as an isotropic medium. A precise understanding of the function of this unusual apparatus—whether as a light guide or as a focusing device—is still lacking. Furthermore, it is also not known whether or not crystalline cone and fiber optics accompanied the lens that is preserved. Conceivably the long prismatic bodies could have encompassed both functions of the lens-crystalline cone assembly. The number of individual optical elements could vary from about a hundred to more than fifteen thousand in a single eye, a range not very different from that of insects. The actual size of the eye, however, often exceeds that of modern arthropods.

Perhaps nothing better than the photographic record can convey a feeling for these structures, and at the same time show the overall shape of the eye and its visual field. Plates 13 and 14 show scanning electron microscope (SEM) pictures (Clarkson 1973c) of some of the ancestral holochroal eyes, from the olenid trilobites of the late Cambrian of Sweden. In plate 13 the eye of *Ctenopyge* (*Mesoctenopyge*) *tumida* is seen partly covered by the corneal membrane. Where this is still present, the swelling caused by the presence of the underlying lenslets is apparent. Where the membrane is removed, only the lensar pits remain of the visual surface. The thin lens profile can be made out by comparing the two regions. Plate 14a shows an intact eye of *Sphaerophthalmus alatus,* and plate 14b the internal mold of an eye of *Sphaerophthalmus humilis*. The external similarities of these primitive eye forms to those of some modern insects (e.g., the ant) is quite remarkable. Plate 15 contains two views of the eye of *Scutellum* (*Paralejurus*) *campaniferum* (Beyrich), a Devonian trilobite from Bohemia. Here the corneal membrane

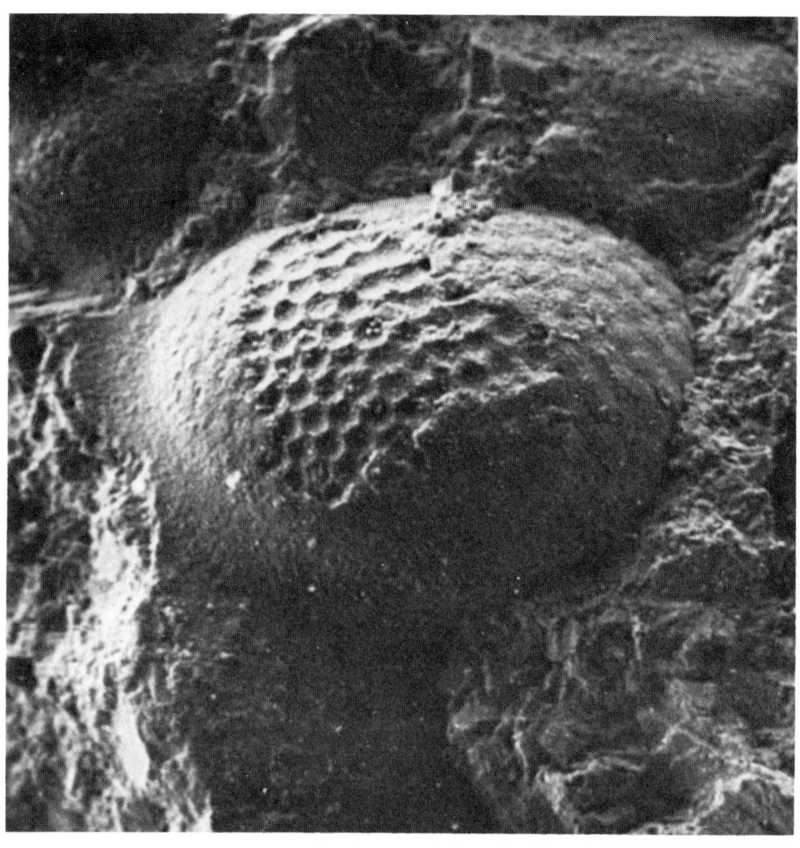

Pl. 13

Plate 13. Scanning electron microscope view of the eye of *Ctenopyge* (*Mesoctenopyge*) *tumida* Westergård (x130). U. Cambrian, Sweden. (Negative loaned by E. N. K. Clarkson; Clarkson 1973c.) The visual surface is partially covered by the corneal membrane.

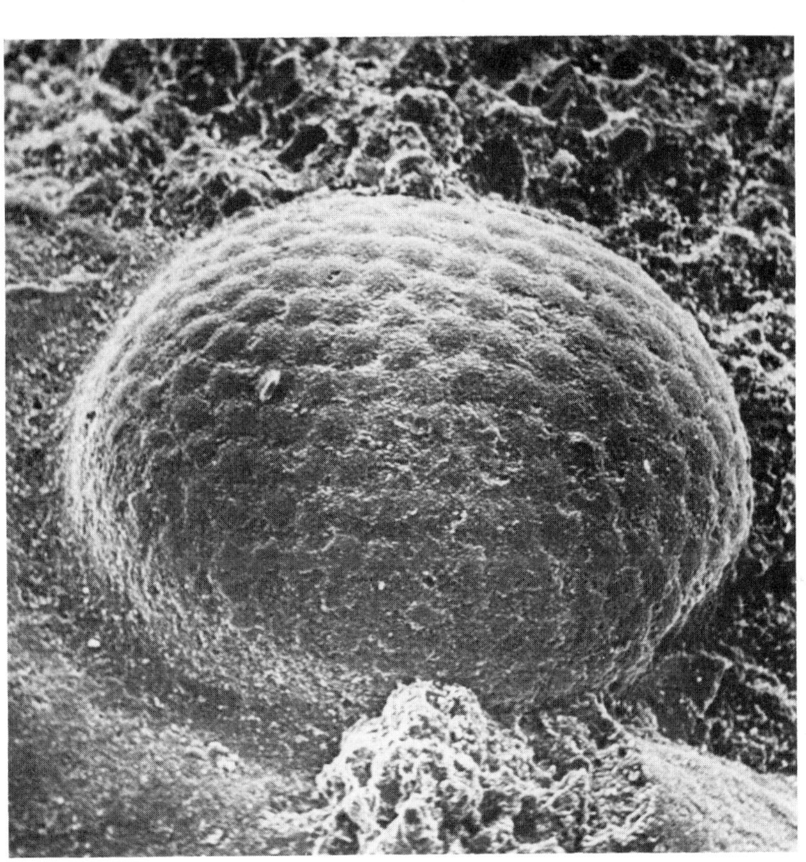

Pl. 14a

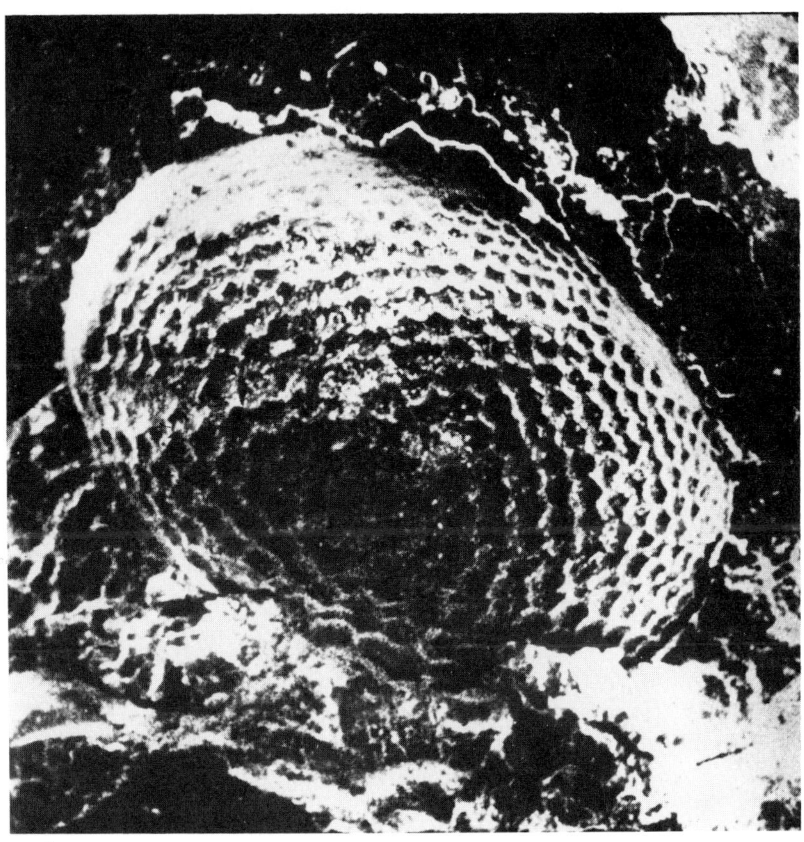

Pl. 14b

Plate 14. (a) SEM view of the eye of a young specimen of *Sphaerophthalmus alatus* (Boeck) (x225). U. Cambrian, Sweden. (Negative loaned by E. N. K. Clarkson; Clarkson 1973c.) The corneal membrane covers the entire visual surface; however, the swellings due to the underlying lenslets are clearly visible. (b) SEM view of an adult eye of *Sphaerophthalmus humilis* (Phyllips) (x100). U. Cambrian, Sweden. (Negative loaned by E. N. K. Clarkson; Clarkson 1973c.) Most of the corneal membrane and lenses are missing, showing the internal mold.

Plate 15. Two views of the same eye of *Scutellum* (*Paralejurus*) *campaniferum* (Beyrich) (x17). (Negative loaned by E. N. K. Clarkson.) This Devonian trilobite from Bohemia, could cover an almost spherical visual field.

Pl. 15a

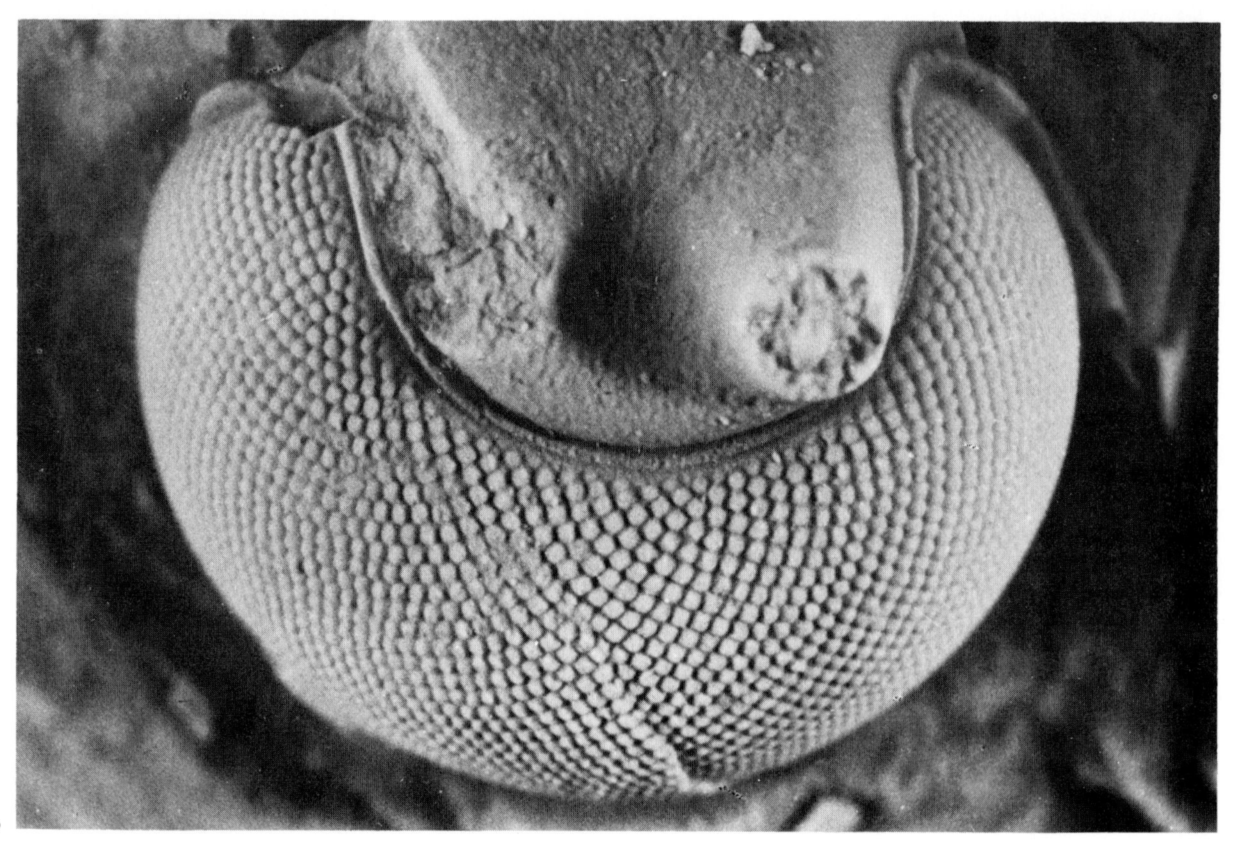

Pl. 15b

is missing and the hexagonal lens pattern is exposed. The toroidal shape of the combined surface of both eyes could evidently span a visual field close to 4π radians. Small gradual deviations from a regular hexagonal disposition of the ommatidia give rise to a spiraling pattern of the dorsoventral files. There are also irregularities and dislocations in the pattern. In Plate 16 we see a SEM picture of a portion of the eye of *Paralejurus brongniarti* (Barrande), also from the Devonian of Bohemia. Although the pattern arrangement is hexagonal, the exposed terminations of the ommatidia appear quite spherical. Plates 17 and 18 show different views of the eyes of *Pricyclopyge binodosa* (Salter), from the Ordovician of Bohemia. The preservation of these remarkable eyes is peculiar since only the ommatidial framework is preserved as an empty dome-shaped beehive. This indicates that between adjacent ommatidia there might have been a wall, which in this case is the only preserved component. The cross section of the ommatidial assembly is exposed in cuts and holes, thus giving a measure of its thickness. The arrangement of the ommatidia is extremely regular. This trilobite had immense eyes in relation to the size of its body. The eyes were placed on the sides of the cephalon, so as to extend to the ventral region, and would almost touch each other in front. In related forms, the eyes actually merged into one uninterrupted visual surface which looked like a panoramic dome. Finally, plates 19a and 19b show two SEM views of a cross section through the visual surface of *Asaphus raniceps* Dalman (Clarkson 1973b). The words of Lindström (1901) in describing this structure are most appropriate here: the lenses "are columnar prisms, like the pillars of basalt." These are the structures made of oriented calcite

Pl. 16

Plate 16. SEM picture of portion of the visual surface of *Paralejurus brongniarti* (Barrande) (x120). Devonian, Bohemia. (Negative loaned by E. N. K. Clarkson.)

Plate 17. Front and side view of the eye of *Pricyclopyge binodosa* (Salter) (x14). Ordovician, Bohemia. (Negative loaned from E. N. K. Clarkson.) In this unusual process of fossilization only the framework of the visual surface is preserved.

Pl. 17a

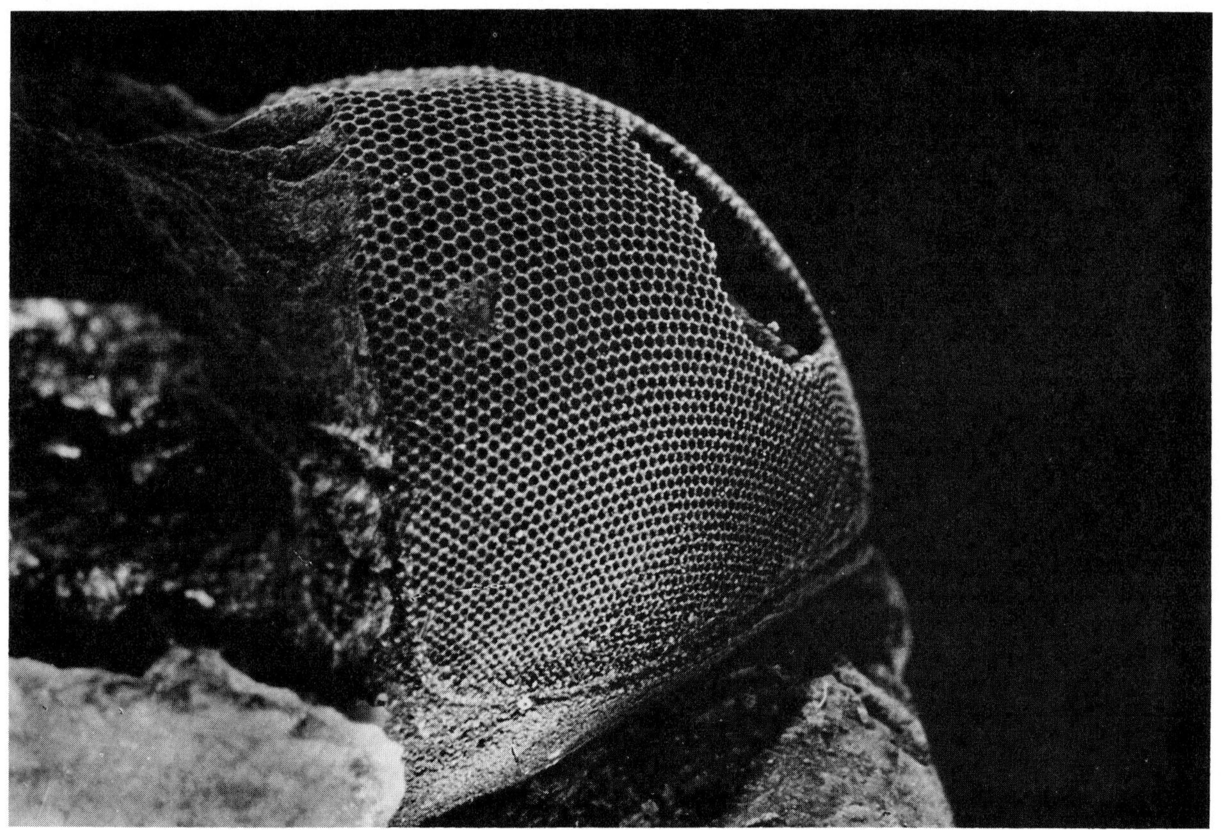

PL. 17b

Plate 18. Top view of the eye of another specimen of *Pricyclopyge binodosa* (Salter), as in plate 17 (x20).

Pl. 18

Pl. 19a

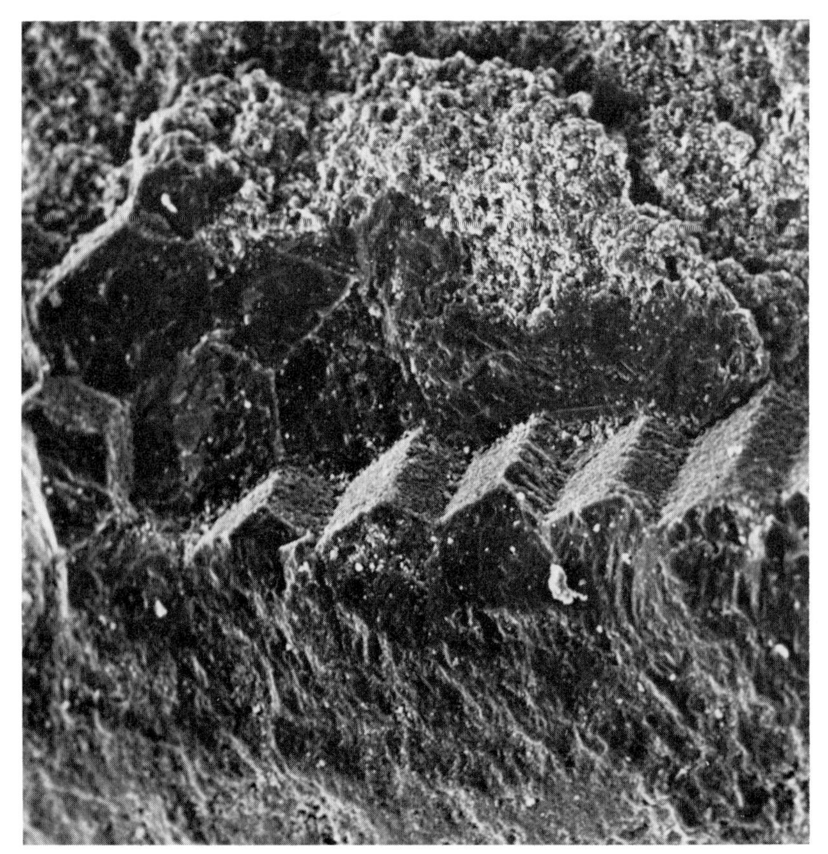

Pl. 19b

Plate 19. SEM views of the prismatic structure of the visual surface of *Asaphus raniceps* Dalman, an Ordovician trilobite from Sweden (a, x170; b, x370). (Negative loaned by E. N. K. Clarkson; (Clarkson 1973c.)

mentioned in connection with the description of figure 7. As noted by Clarkson, the symmetrically radiating arrangement of the ommatidia in this trilobite reminds one of the arrangement of the superposition eye of modern arthropods. For illustration of this point, figure 8 shows a cross section through the eye of *Asaphus* compared with that of a night moth, *Deilephila elpenor* (in the light adapted condition, in order to show the crystalline cones) (Höglund 1965).

3.3.4. Schizochroal Eyes

The lenses in aggregate, or schizochroal, eyes are generally larger than those of holochroal eyes. They are also less numerous, ranging from a few to several hundred in each lateral eye. This type of eye is characteristic of the phacopid trilobites and a few others. It appears with Ordovician species and persists until trilobite extinction. A particular form of evolutionary optimization of the optical function has taken place in the schizochroal eyes, to make them perhaps the most sophisticated dioptric elements ever produced by nature.

The lenses of phacopid trilobites were doublet structures built to correct the otherwise large spherical aberration of simple thick lenses. The evidence for this feat of trilobites is so extraordinary that it deserves a chronological description. The story begins when I met Dr. E. N. K. Clarkson at the Oslo International Conference on Trilobites in July 1973. Following my presentation of the evidence concerning optimal light collection properties in the *Limulus* ommatidia, Dr. Clarkson and I sat for coffee to talk about trilobite eyes. On that occasion I learned from Dr. Clarkson of his studies of the lens structure in the phacopid trilobites. Although Lindström (1901) had already pointed out that the lenses of schizochroal eyes seemed to have a peculiar substructure, it was not until after a systematic study by Dr. Clarkson in 1968 that such substructure could be established in its details. Coming back to Chicago with some napkin sketches of doublet lenses split by a curious wavy surface, I was convinced that trilobites must have tried to correct their optics. The real surprise came when I found out *how* the little animals coped with the problem. This discovery happened when I was browsing through the sacred works of Christian Huygens, the father of wave optics. In his *Traité de la Lumière,* published in 1690, Huygens described the construction of an aspherical aplanatic lens, the outline of which resembled unmistakeably the wavy shape seen in Clarkson's sketches. This is not all: the mention by Huygens of similar earlier results by Descartes led me to peruse his *La Géometrie,* published in 1637. There I found a second construction, somewhat different from that of Huygens, but designed to perform the same function. This matched a second version of trilobite lens shapes described by Clarkson. Armed with the conviction that trilobites had solved a very elegant physical problem and apparently knew about Fermat's principle, Abbe's sine law, Snell's laws of refraction, and the optics of birefringent crystals, I set out to inform Dr. Clarkson of the meaning of his trilobites' lens shapes. While a complete understanding of the marvelous phacopid lenses is still the object of our research, I cannot omit in this context a description of some of the preliminary and yet unpublished conclusions of this work.

Figure 9 reproduces a page of the *Treatise on Light* by Christian Huygens which contains the construction of an aspherical aplanatic lens. Shown in the bottom portion of the figure is the reconstruction of the lens

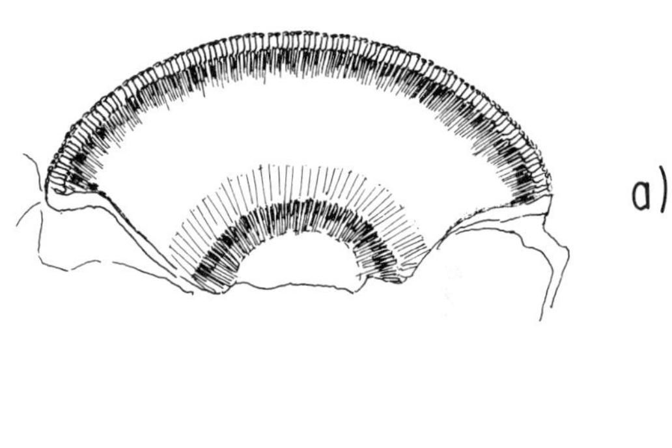

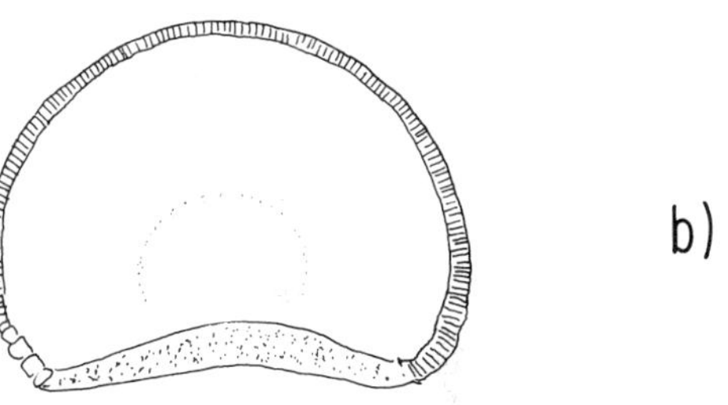

Fig. 8

Fig. 8. (a) Cross sectional view of the superposition eye of the night moth, in the light-adapted condition (adapted from Höglund 1965). (b) Cross-sectional view of the holochroal eye of Asaphus (adapted from Clarkson 1973b). The ommatidial arrangement was probably similar in the two cases.

ON LIGHT. CHAP. VI 117

such that the path of the light from the point L to the surface AK, and from thence to the surface BDK, and from thence to the point F, shall be traversed everywhere in equal times, and in each case in a time equal to that which the light employs to pass along the straight line LF of which the part AB is within the glass.

Let LG be a ray falling on the arc AK. Its refraction GV will be given by means of the tangent which will be drawn at the point G. Now in GV the point D must be found such that FD together with $\frac{3}{2}$ of DG and the straight line GL,

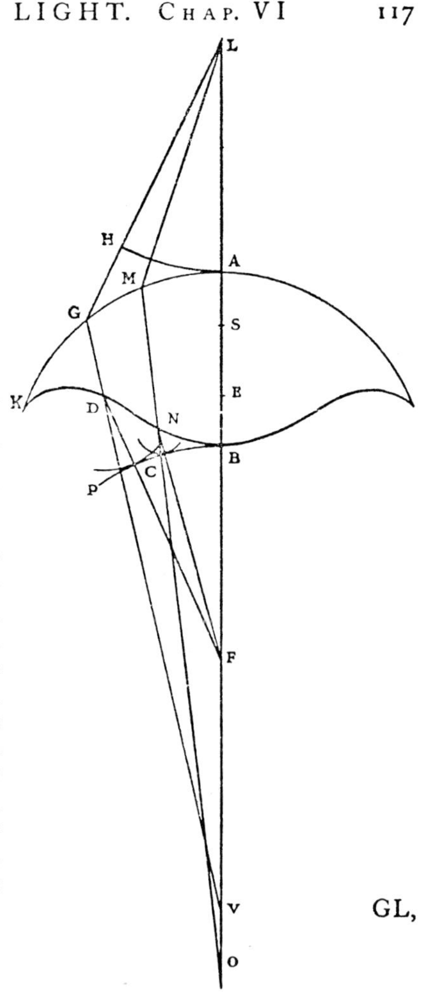

Fig. 9a

438 ŒUVRES DE DESCARTES. 366-367.

effect que le precedent, & que la conuexité de l'vne de ſes ſuperficies ait la proportion donnée auec celle de l'autre.

entre les lignes AM & YM, & qu'il faille trouuer la figure du verre ACY, qui face que tous les rayons qui vienent du point G s'aſſemblent au point F.

On peut de rechef icy ſe ſeruir de deux ouales, dont l'vne, AC, ait G & H pour ſes poins bruſlans, & l'autre, CY, ait F & H pour les ſiens. Et pour les trouuer, premierement, ſuppoſant le point H, qui eſt commun a toutes deux, eſtre connu, ie cherche AM par les trois poins G, C, H, en la façon tout maintenent expliquée : a ſçauoir, prenant k pour la difference qui eſt

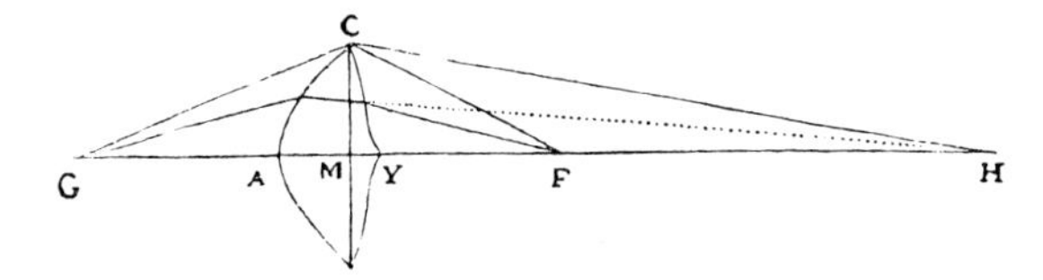

entre CH & HM, & g pour celle qui eſt entre GC & GM; & AC eſtant la premiere partie de l'ouale du premier genre, i'ay $\frac{ge+dk}{d-e}$ pour AM. Puis ie cherche auſſy MY par les trois poins F, C, H, en ſorte que CY ſoit la premiere partie d'vne ouale du troiſieſme genre : & prenant y pour MY, & f pour la difference qui eſt entre CF & FM, i'ay $f+y$ pour celle qui eſt entre CF & FY : puis, ayant deſia k pour celle qui eſt entre CH & HM, i'ay $k+y$ pour celle qui eſt entre CH & HY, que ie ſçay deuoir eſtre a $f+y$ comme e eſt a d, a cauſe de l'ouale du troiſieſme genre. D'où ie trouue que y ou MY eſt $\frac{fe-dk}{d-e}$; puis, ioignant enſemble les deux quantités trouuées pour AM & MY, ie trouue $\frac{ge+fe}{d-e}$ pour la toute AY. D'où il ſuit que, de quelque coſté que ſoit ſuppoſé le point H, cete ligne AY eſt touſ-

Fig. 10a

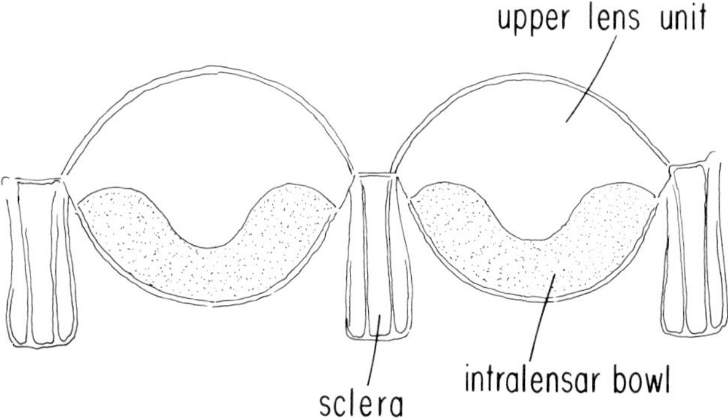

Fig. 9b

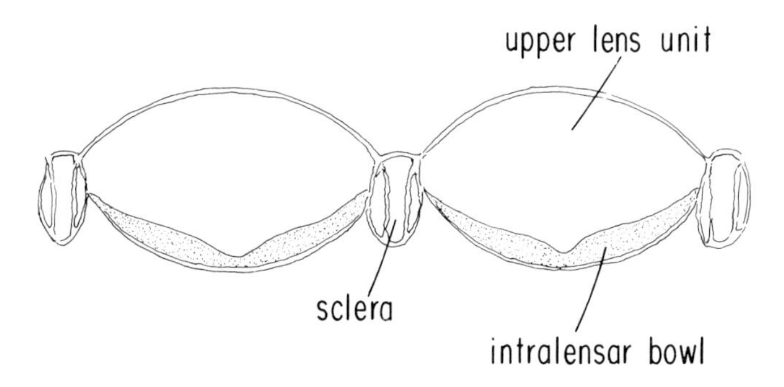

Fig. 10b

Fig. 9. (a) Construction of a lens free of spherical aberration, from C. Huygens' *Treatise on Light* (courtesy of the University of Chicago Press). (b) Cross sectional view of the lenses in the eye of *Crazonaspis struvei* Henry, an Ordovician trilobite from Brittany (Clarkson 1968). The intermediate surface is shaped accordingly to the prescription by Huygens.

Fig. 10. (a) Construction, similar to that of figure 9, after Descartes in *La Géometrie*. Here again, the shape of the second surface makes the lens free of spherical aberration. (b) Cross-sectional view of the lens structure in the eye of *Dalmanitina socialis* (Barrande) (Clarkson 1968). The intermediate surface is shaped in remarkable accord with the design by Descartes.

Fig. 11. Ray tracing through the lens of *Crozonaspis struvei* Henry. On the right of the optical axis, the internal lens structure is ignored. Very large spherical aberration ensues for any choice of the refraction index of the lens (n = 1.66 for the construction indicated). When the internal structure is taken into account, on the left-hand side of the axis, correction of spherical aberration obtains for the combination of refractive indices indicated. This suggests that the two lens elements were made of oriented calcite and chitin respectively.

structure in *Crozonaspis struvei* Henry, an Ordovician trilobite from Brittany (Clarkson 1968). The structure is characterized by a lens unit, having the Huygens shape for its second surface, coupled with an *intralensar bowl* which completes an overall biconvex shape. Figure 10 shows a similar comparison between the construction by Descartes (from *La Géometrie*) and the lens structure of *Dalmanitina socialis* (Barrande), another phacopid trilobite (Clarkson 1968). In both Huygens' and Descartes' constructions the requirement is that, given the profile of the first refracting lens surface and the index of refraction of the lens, all rays from a point in object space should be made to converge to a point in image space. The construction determines the shape of the second surface, one of a class of surfaces called Cartesian Ovals.

As can be read from Huygens' words in figure 9, the above requirement is met when all rays from the point source to the point image, through the lens, traverse their path in equal times. In modern language, all rays from the source will converge to a point image if they traverse minimal and identical "optical paths." A lens which satisfies this condition is free of spherical aberration and is called aplanatic. The aspheric cartesian surface which results is described by a fourth-degree equation. Why are Huygens' and Descartes' lens shapes different? Huygens imposes the thickness of the lens while Descartes does not. If a lens shaped like the upper unit of the trilobite lens is already sufficient to eliminate spherical aberration, what then is the purpose of the intralensar bowl? The answer was given by ray-tracing through a lens system like that reproduced in figure 9. This construction is shown in figure 11. The indices of refraction of the lens parts being unknown, they must

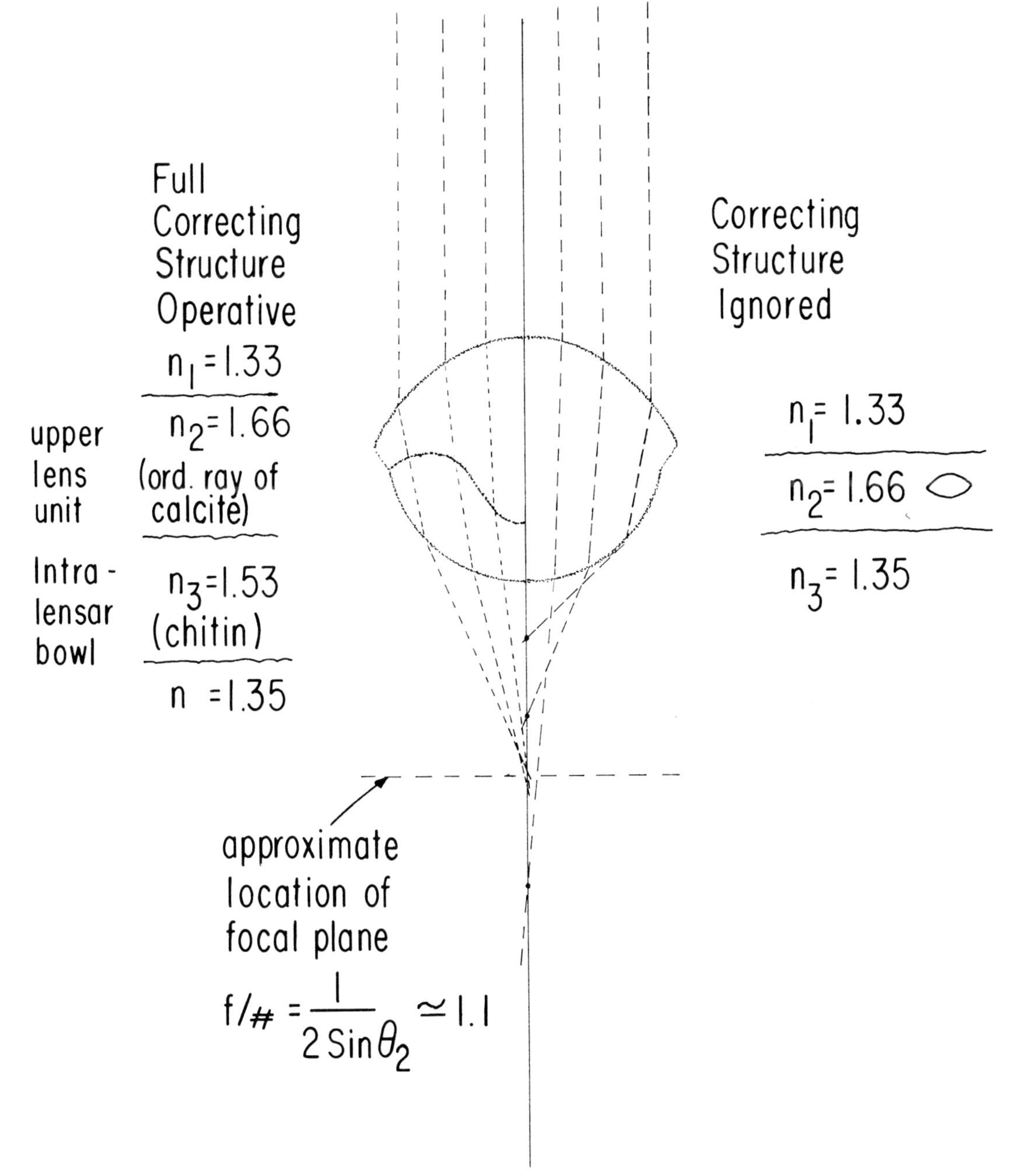

Fig. 11

Plate 20. Frontal view of the left eye of *Dalmanites pratteni* Roy, a Devonian trilobite from Illinois (x14). (Specimen loaned by the Field Museum of Natural History, Chicago.)

Pl. 20

Pl. 21

Plate 21. Frontal view of the right eye of *Dalmanites pratteni* Roy, as in plate 20. The lens structure described in the text is exposed in several instances, where the front lens-unit has split away. In some elements the entire lens appears intact, encased in the cylindrical sclera. Among schizochroal eyes, the eyes of this trilobite exhibit the largest number of lenses ever recorded.

be guessed by trial and error, with the knowledge, however, that there must be water on the outside of the lens, and some kind of organic fluid on the inside. There is in fact only one choice of indices for which the lens brings an incident parallel beam to a focus. This involves the upper lens unit being made of oriented calcite (refracting index $n = 1.66$ on axis) and the intralensar bowl being made of chitin, for example (refracting index $n = 1.53$). The problem is solved. There is independent evidence to show that the lenses of phacopid trilobites were made of oriented calcite, *in vivo* (Towe 1973). The above result, therefore, seems reasonable. The role of the intralensar bowl is also understood. It is needed for convergence to occur throughout the lens. In the constructions by Huygens and Descartes, their lenses were made of glass in air. Here we are dealing with calcite ($n = 1.66$ for the ordinary rays) immersed in water ($n = 1.33$). The upper unit alone cannot function as desired in this condition. In fact the more peripheral rays would diverge. However, the intralensar bowl restores the focusing property. In other words, in our case the Huygens or Descartes upper lens acts as an actual correcting element of an otherwise very poor lens, as the Schmidt plate does in modern telescopes. On the right-hand side of figure 11 we see how the lens would behave if it were made of a solid unit, without corrector. The spherical aberration would prevent the formation of any sharp image. Other problems still have to be solved, in particular how to dispose of the ghost image which would be formed when rays impinge on the birefringent upper unit, off axis. The basic outline of the lens behavior, however, is deciphered.

The feat of trilobites in optimizing their optical apparatus raises very relevant questions as to why such perfection was needed. Most likely the phacopid trilobites developed very thick lenses for the purpose of increasing their light collecting efficiency. For thick uncorrected lenses however, the collected light would be diffused along the axis due to the spherical aberration without reaching the intensity needed for firing the optic neurons. With defect correction, the thick lenses would perform the required function, that of concentrating light so that its intensity exceeded a certain minimum detection threshold. As suggested by Clarkson, this may indicate that thick-lensed trilobites were nocturnal or crepuscular animals, or that they lived in very turbid waters.

In the actual fossilized eye of phacopids, the evidence for the extraordinary doublet structure is only occasionally obvious, when, due to differential mineralization, the intralensar bowl splits from the upper lens unit. More often the structure can only be seen in polished sections of the lens surface, and in many fossils it is missing altogether even if present in the original eye. A fortunate specimen in this respect was located by the author at the Field Museum of Natural History in Chicago. It is the holotype of *Dalmanites pratteni* Roy (Roy 1933), a spectacular phacopid trilobite from the Devonian of Illinois. Front views of the right and left eye of this rare trilobite are shown in plates 20 and 21 respectively. Each eye contains more than 770 lenses, an absolute record for schizochroal eyes. Several of the lenses are split and show the intralensar bowl. In other cases the upper lens unit is still in place: and in still others the entire doublet is missing and the cavity left is the lensar pit. A detailed study of the kind of doublet structure present here is still in progress.

The Huygens' type lens substructure has been established by Clarkson (Clarkson 1969), for the eye of *Reedops sternbergi* (Hawle and Corda), represented in plate 22. The evidence, however, is accessible only through sectioning. Two SEM views of the visual surface of this trilobite are shown in plates 22b and 23a. Several other schizochroal eyes are shown next. Plate 23b represents the eye of *Dalmanites verrucosus* Hall, of the Silurian Waldron shale. In a remarkable condition of preservation are the eyes of *Phacops rana crassituberculata* Stumm, from the Devonian Silica shale of Ohio, shown in plates 24 and 25. These eyes show clearly a band of "sensory fossettes" below the visual surface. The lenses are quite deeply set inside the sclera. A variant of this trilobite, also from the Silica shale, is *Phacops rana milleri* Stewart, whose eye is shown in plate 26. In contains more lenses than the previous one (see, e.g., Eldredge 1972). The eye of *Eophacops trapeziceps* (Barrande) is shown in plate 27, and that of *Chasmops odini* (Eichwald) in plate 28. The latter is an Ordovician trilobite from Estonia. In most phacopid eyes, the field of view of each eye could span 180° longitudinally, and a strip of 10–20 degrees latitudinally. The schizochroal eye with the smallest number of lenses is shown in the SEM photograph on plate 29. It is the eye of *Denckmannites volborthi* (Barrande) of the Devonian of Bohemia. The schizochroal eye disappeared toward the end of the Devonian, with the extinction of the phacopid trilobites. It is unfortunate that the genetic information of such a perfected visual apparatus became lost to further evolution in the animal kingdom.

Pl. 22a

Plate 22. (a) The eye of *Reedops sternbergi* (Hawle and Corda), a Devonian trilobite from Bohemia (x22). (Negative loaned by E. N. K. Clarkson; Clarkson, 1969.) Sectioning of the lenses of similar specimens by Clarkson has revealed an internal structure similar to that described in figure 9. (b) SEM view of portion of the eye of *Reedops sternbergi*, as in (a) (x120). (Negative loaned by E. N. K. Clarkson.)

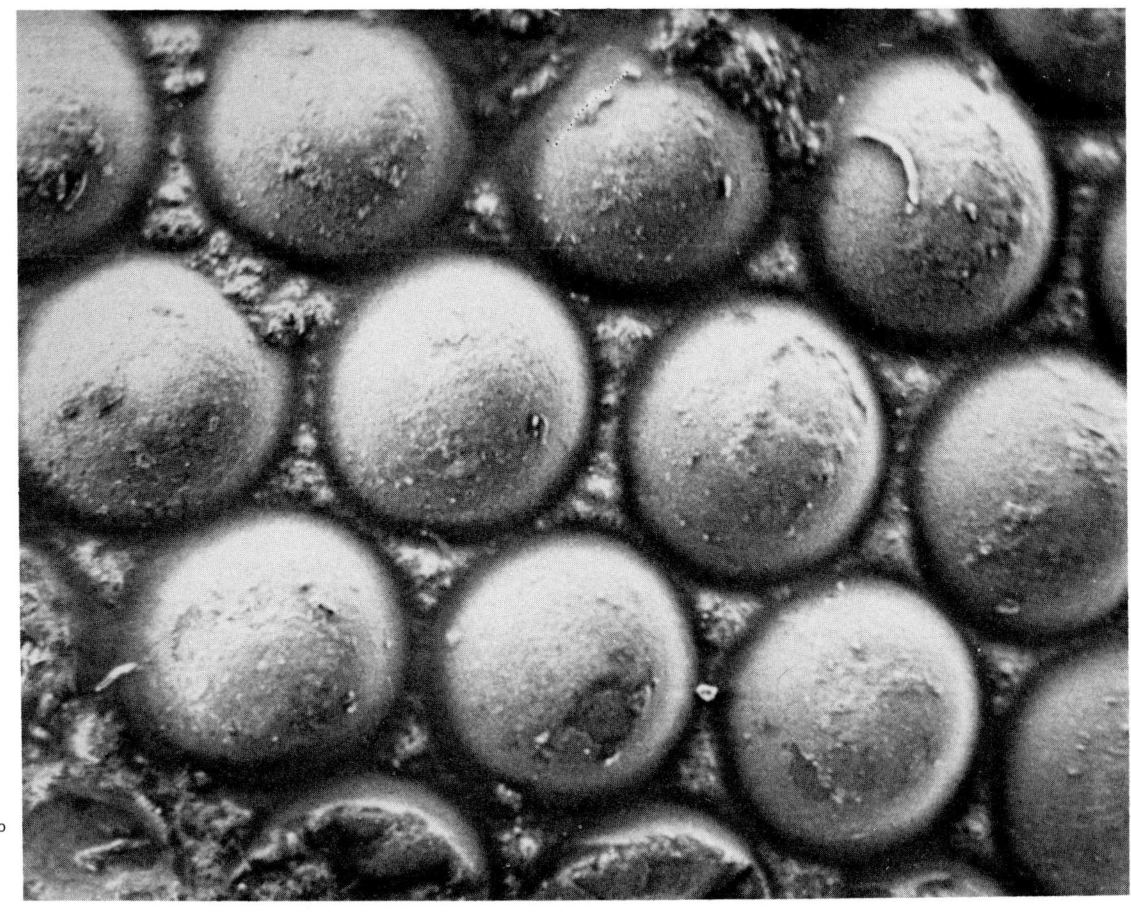

Pl. 22b

Plate 23. (a) Another SEM view, at a glancing angle, of the lenses of *Reedops sternbergi* as in plate 22 a and b (x240). (b) Right eye of *Dalmanites verrucosus* Hall, a Silurian phacopid trilobite from the Waldron shale, Waldron, Indiana (x10). (Loaned from the Field Museum of Natural History, Chicago.) The visual surface is still partially immersed in the shale matrix.

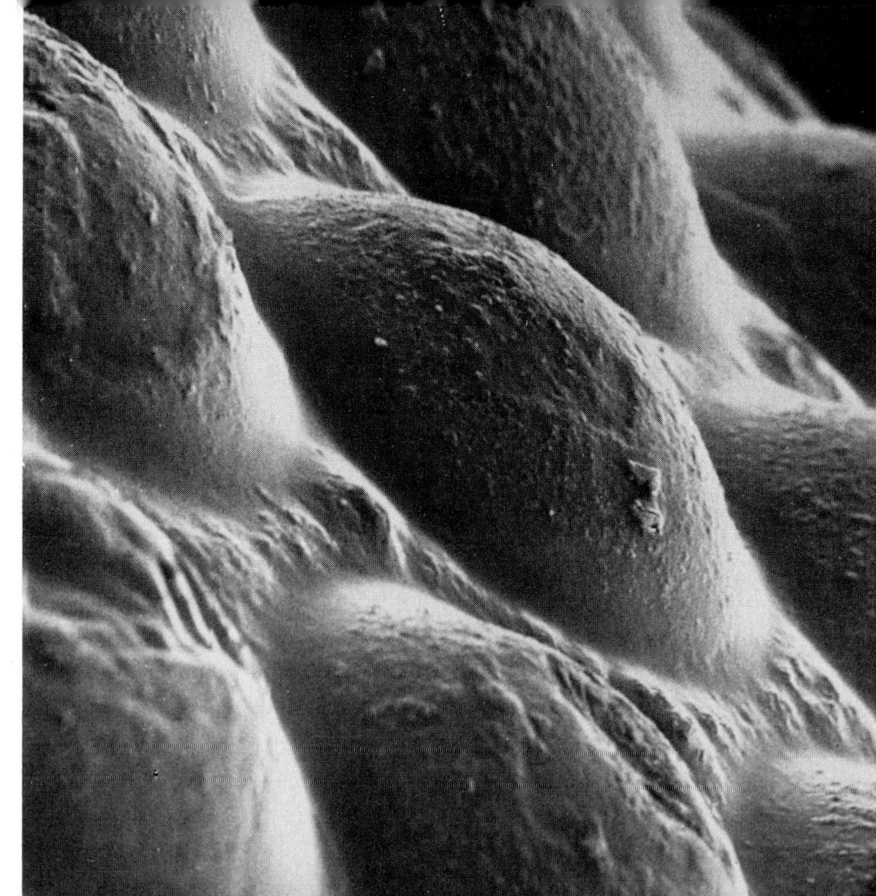

Pl. 23a

Pl. 23b

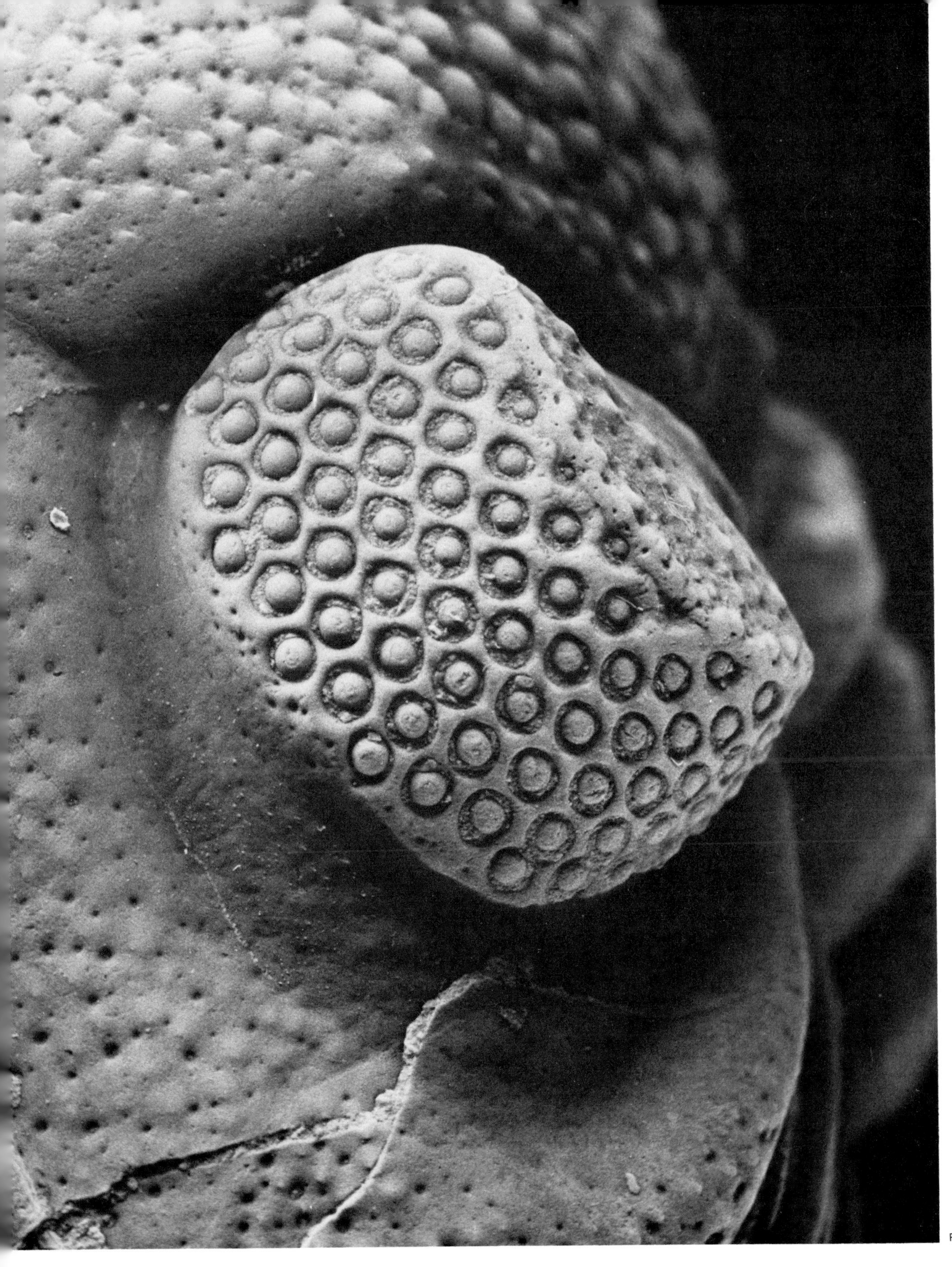

Plate 24. Left eye of *Phacops rana crassituberculata* Stumm, from the Devonian Silica shale at Sylvania, Ohio (x14). (RLS coll.) Specimen whitened with magnesium oxide. New lenses are added at the top of the visual surface and appear incompletely developed.

Pl. 24

Plate 25. Right eye of another specimen of *Phacops rana crassituberculata* Stumm, as in plate 24 (x14). (RLS coll.) Specimen whitened with magnesium oxide. The lenses are deeply encased in the scleral surface.

Pl. 25

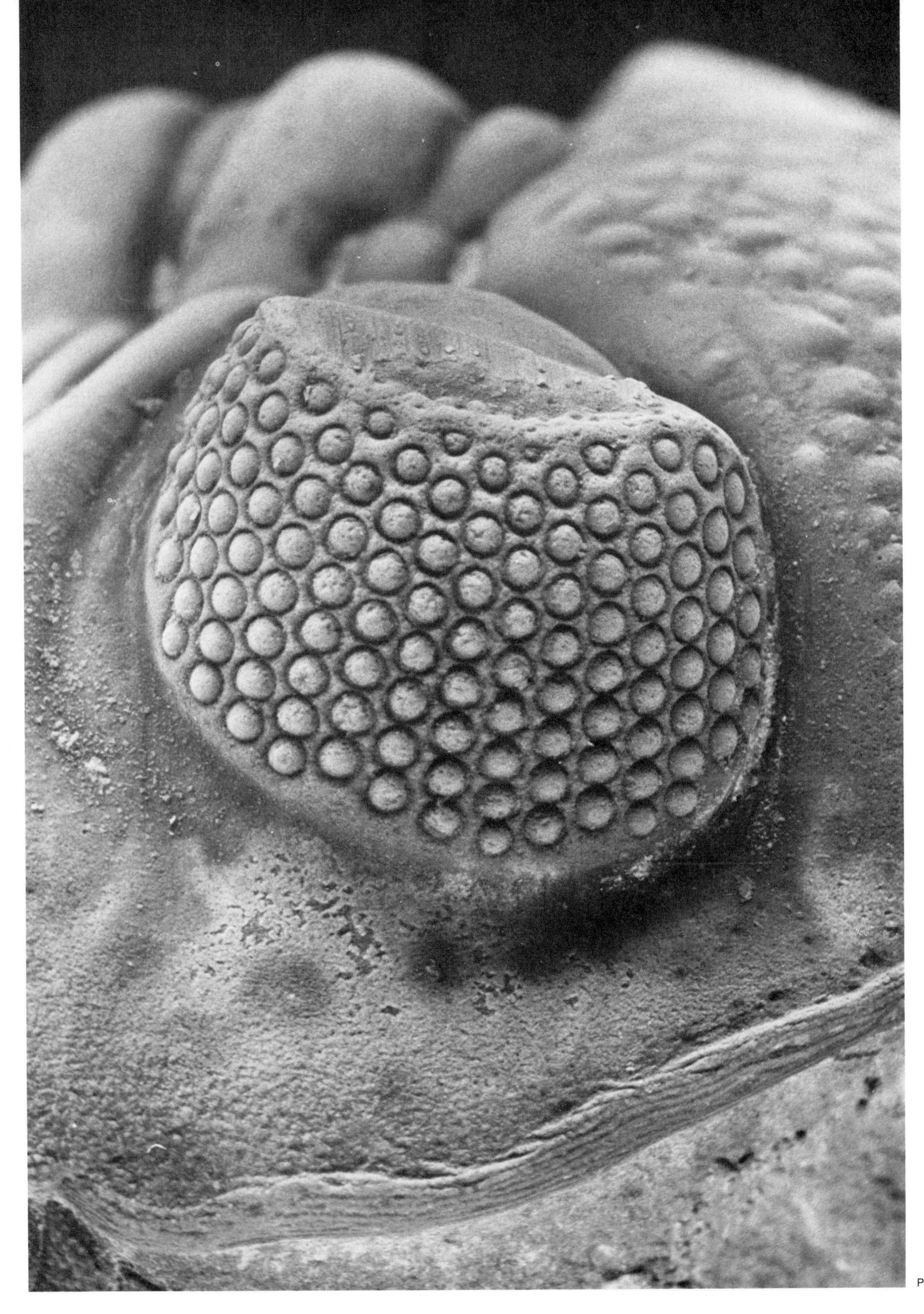

Plate 26. The right eye of *Phacops rana milleri* Stewart, from the Devonian Silica shale of Sylvania, Ohio (x 15.) (RLS coll.) This trilobite, familiar to many collectors, differs from the variety shown in plates 24 and 25, mostly in the larger number of eye lenses and in the less tuberculate surface of the palpebral lobe.

Pl. 26

Pl. 27

Plate 27. Right eye of *Eophacops trapeziceps* (Barrande), a Silurian trilobite from Bohemia. (x20.5). (Negative loaned by E. N. K. Clarkson.) The lenses here are very prominent above the plane of the sclera.

Pl. 28

Plate 28. Right eye of *Chasmops odini* (Eichwald), Ordovician trilobite from Estonia (x21). (Negative loaned by E. N. K. Clarkson; Clarkson 1966.)

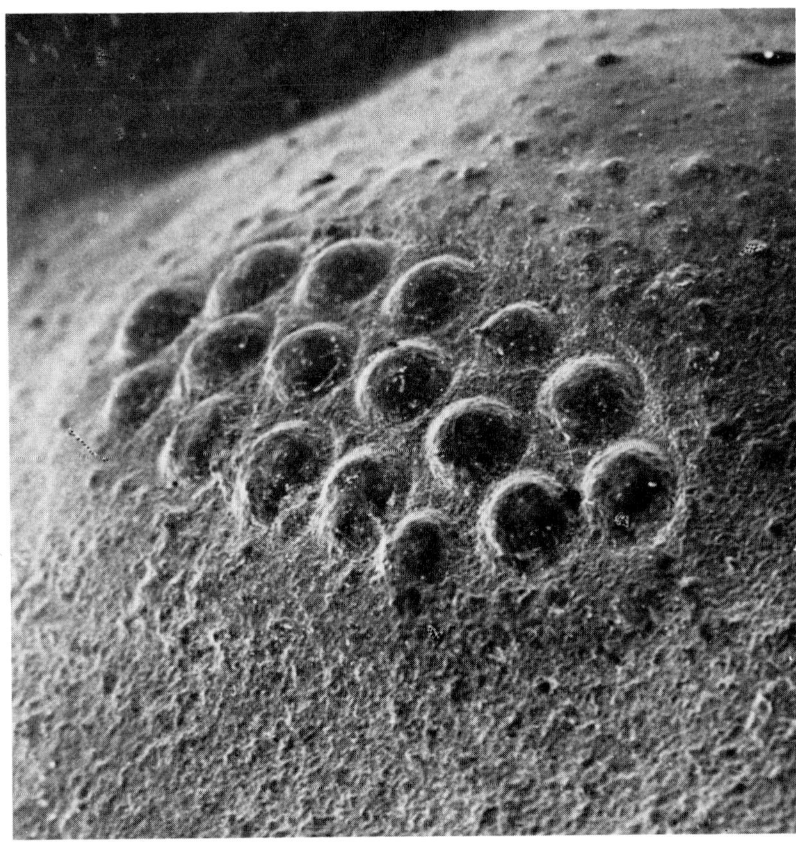

Pl. 29

Plate 29. SEM view of the eye of *Denkmannites volborthi* (Barrande). Devonian of Bohemia (x72). (Negative loaned by E. N. K. Clarkson.) This type of schizochroal eye is among those containing the smallest number of lenses.

3.4. Enrollment

The great majority of trilobites could roll themselves up so that only the hard carapace was then exposed. In this condition, various types of spines which may have adorned the exoskeleton became functional, protruding from the enrolled body in a defensive role. There are many examples of this behavior in modern arthropods. A garden variety millipede (*Sphaerotherium*) is often seen rolled up. In the horseshoe crab this function is only partially available, since the thorax is fused with the cephalon. The articulation of the abdominal region, then, allows only a partial enrollment, drawing up the powerful telson to make a right angle with the rest of the body. Earlier related forms, however, could double up completely.

The mechanism of enrollment in trilobites is characteristic of particular phylogenetically related groups, as was recently emphasized by Bergström (1973a). The thoracic segments were clearly engineered to permit this function, and a variety of articulating joints have been identified. To give a most schematic description of the thorax, individual segments were hinged at two points, often characterized by visible notches, proximal to the axial furrows. Rotation of the segments around these pivot points would separate the axial rings, exposing the articulating half rings, and at the same time bring the pleural extremities to overlap each other. Special devices, called *panderian organs,* would, like a doorstep, limit the extent of enrollment. Other devices, the *vincular furrows,* would ensure a safe interlocking of the pygidial and cephalic margins, as if to discourage casual intruders.

A few major modes of enrollment have been recognized by Bergström in his recent review (Bergström 1973a). These differ from

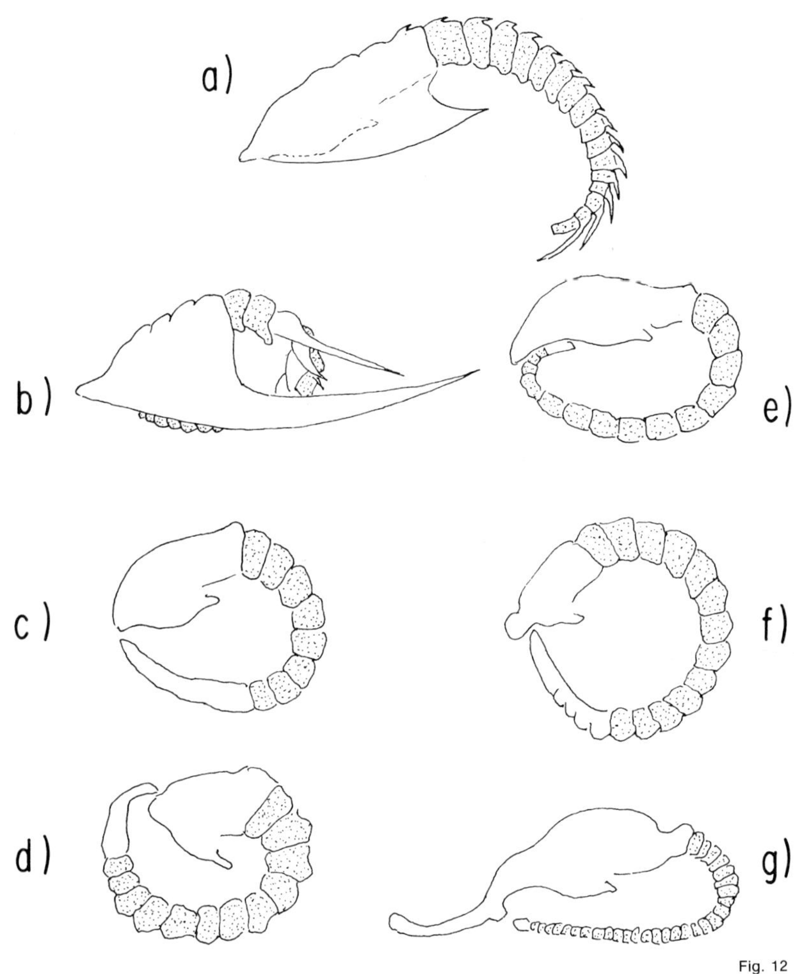

Fig. 12. Schematization of the types of enrollment in trilobites. (a) Incomplete enrollment, as in *Kjerulfia*. (b) Cylindrical enrollment as in *Fallotaspis*. (c) Spheroidal enrollment proper, as in *Asaphus*. (e) Spiral enrollment proper, as in *Ellipsocephalus*. (f) Uncoiled spiral enrollment, as in *Flexicalymene*. (g) Basket and lid enrollment, as in *Harpes*. Adapted from Bergström 1973a.

previous descriptions in that they reflect functional characteristics rather than purely morphological distinctions; these forms of enrollment are sketched in figure 12. Some trilobites were unable to roll up completely, and this mode is called *incomplete* enrollment. Of those which could roll up completely, a distinction is made between *spheroidal* and *spiral* enrollment. Each contains a sequence of subcases. In the spheroidal enrollment the pygidium comes to rest with its ventral side in contact with the cephalic doublure, not inside it, and the pleurae close the exoskeletal basket laterally. When the pleurae do not wrap around to seal the basket laterally, the enrollment is called *cylindrical*. When the pygidial termination overlaps the cephalic margin the mode is called *inverted spiral enrollment* but is still considered part of the spheroidal main group. The spiral enrollment series contains the case in which the *dorsal* part of the pygidium contacts the ventral side of the cephalon; the *uncoiled spiral enrollment,* characteristic of the calymenids, where the pygidium is fully visible even in the enrolled condition; and finally the so-called *basket and lid* enrollment. Both spheroidal and spiral enrollment seem to have evolved from the incomplete enrollment of early Cambrian trilobites. Although enrollment most likely represented a defense mechanism, in the long run it may have precipitated the disappearance of the trilobites. It is conceivable that, with the advent of fishes, an enrolled trilobite could have been swallowed more easily than an outstretched one. Fortunately for us, if not for them, enrolled trilobites made better fossils, since the hard carapace is often quite impervious to weathering agents.

The Middle Cambrian trilobite *Elrathia kingii* (Meek) from the Wheeler formation of

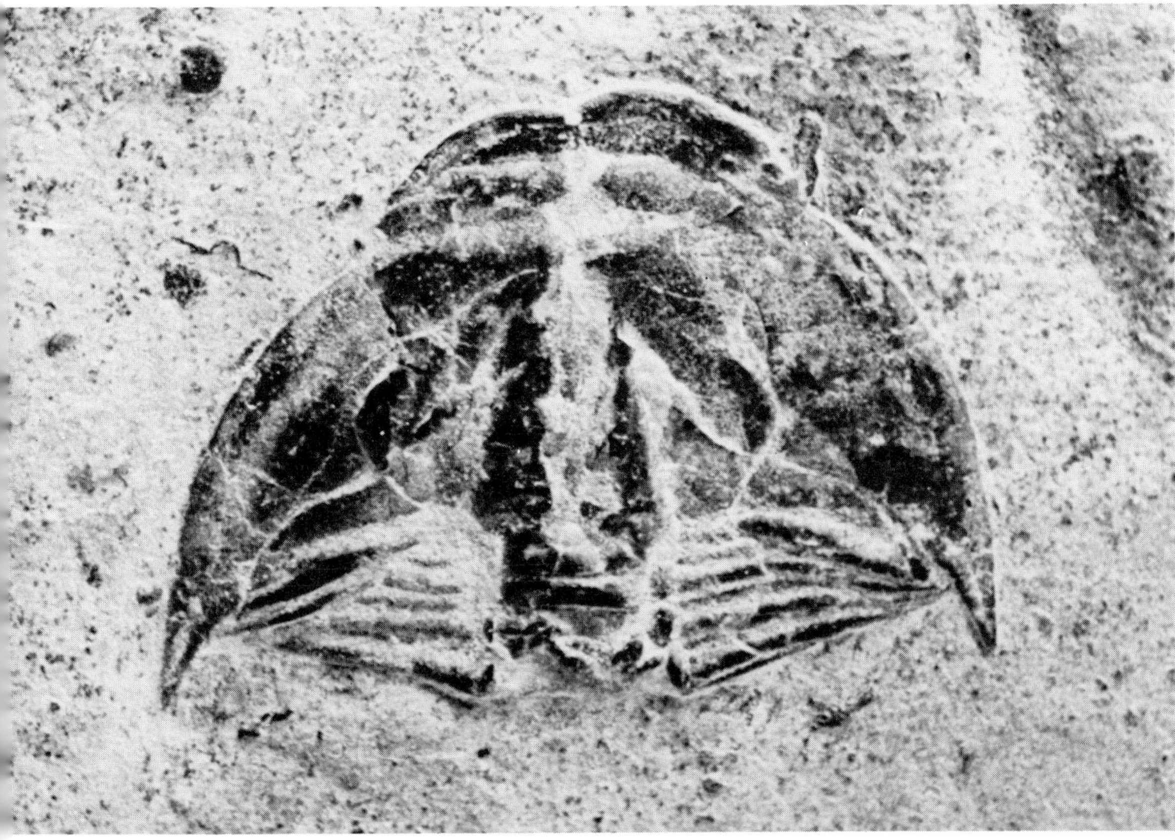

Pl. 30

Pl. 31

Plate 30. Enrolled specimen of *Elrathia kingii* (Meek), seen from the cephalic side. M. Cambrian, Wheeler formation, Utah (x10). (RLS coll.) This is an early example of spheroidal enrollment.

Plate 31. A perfect example of spheroidal enrollment. The trilobite is *Encrinurus variolaris* (Brongniart). Silurian, Dudley, England (x7.5). (Negative loaned by E. N. K. Clarkson.)

Utah offers one of the earliest examples of spheroidal enrollment. In plate 30 we see an enrolled specimen of this species from the cephalic side. The ventral side of the pygidium is seen to protrude beyond the cephalic border, possibly as a result of compression. The entire specimen is in fact considerably flattened. This picture shows how the genal spines protrude from the body when the trilobite is enrolled.

Another case of spheroidal enrollment, this time in full undistorted relief, is shown in plate 31 (Clarkson 1973a). The trilobite is *Encrinurus variolaris* (Brongniart), from the Silurian of Dudley, England. Here the closure of the basket is perfect, and one begins to understand why the outline of a trilobite's carapace is often so longitudinally symmetrical. The two halves have to match rather accurately to ensure a complete enclosure in the enrolled condition.

In Plate 32a, b, c, and d we see four different moments in the enrollment process of *Phacops rana*. In the Devonian Silica shale at Sylvania, Ohio, this trilobite is commonly found in various postures, from the outstretched to the completely enrolled one. This has nothing to do with incomplete enrollment. The mode describes the ultimate capacity and not the intermediate steps to reach it. Although the character in 32d did not quite perform according to the prescription, we are still dealing with a classical example of spheroidal enrollment.

In plates 33 and 34 we encounter the same *Dalmanites pratteni* Roy which in section 3.3 was shown to have such special eyes. The specimen is in fact perfectly enrolled and represents an example of spheroidal enrollment in which the pygidium extends quite a way beyond the cephalic border. It must be realized that, due to its construction, the trilobite could not do any

Plate 32. Various postures in the process of enrollment of *Phacops rana*, from the Devonian Silica shale of Sylvania, Ohio (x2.7). (RLS coll.)

Plate 33. The trilobite represented in enrolled position is the same specimen of *Dalmanites pratteni* Roy, whose eyes are shown in plates 20 and 21 (x2.5). Loaned by the Field Museum of Natural History, Chicago. In this photograph the portion of matrix covering the cephalon and carrying the tip of the pygidium is shown removed from the enrolled trilobite. The cephalic margin is seen to fit tightly against the ventral side of the pygidium.

Pl. 33

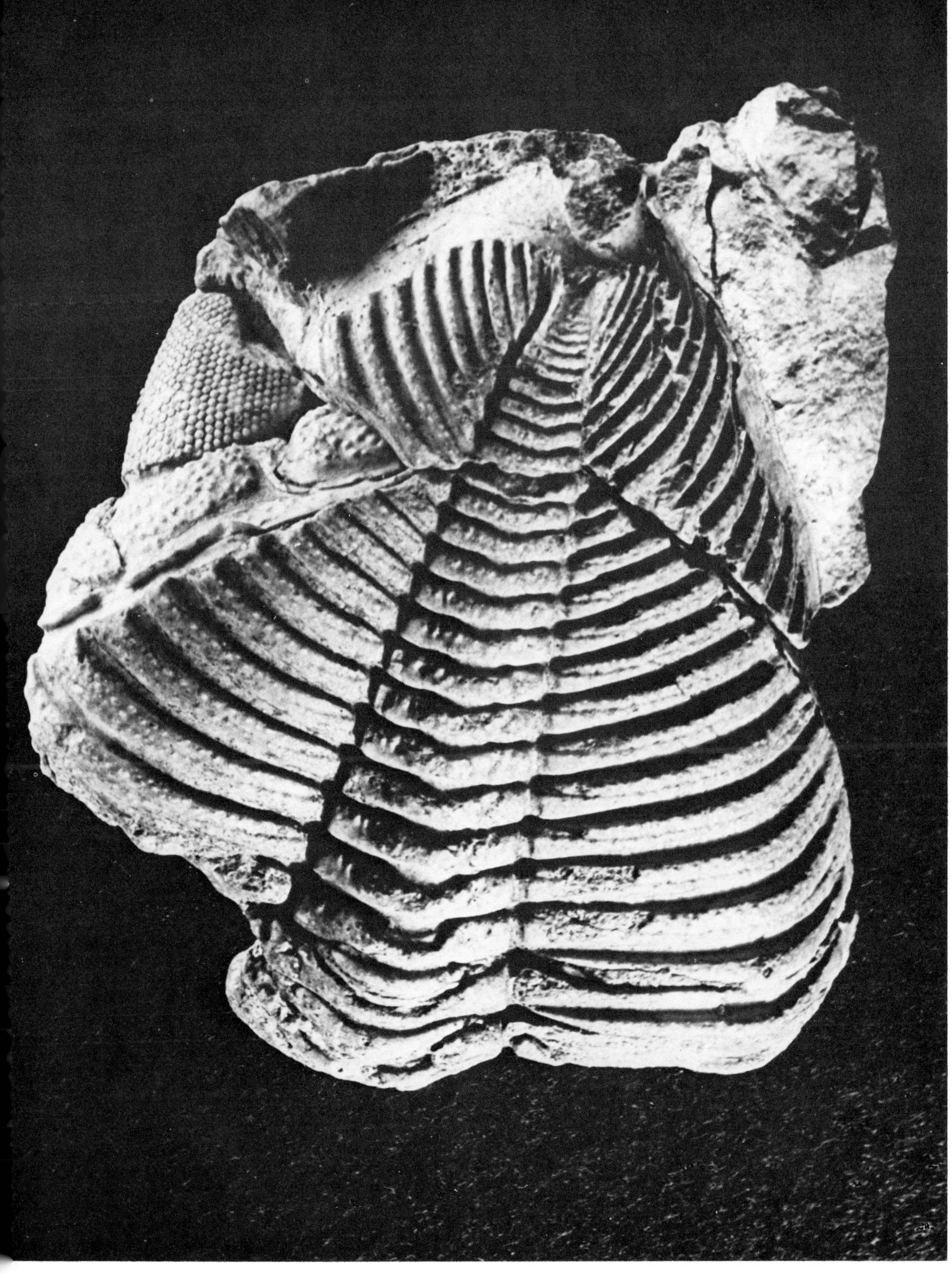

Plate 34. The two parts of the specimen in plate 33 are now assembled to show how the trilobite was originally found. In no way could the trilobite bring the cephalic margin to match the border of the long pygidium. Even so, enrollment provided effective protection, also in view of the presence of a robust telson, not preserved here, protruding from the tip of the pygidium.

Pl. 34

Plate 35. A perfectly enrolled specimen of the Ordovician trilobite *Flexicalymene meeki* (Foerste) from Waynesville, Ohio (x10). (RLS coll.)

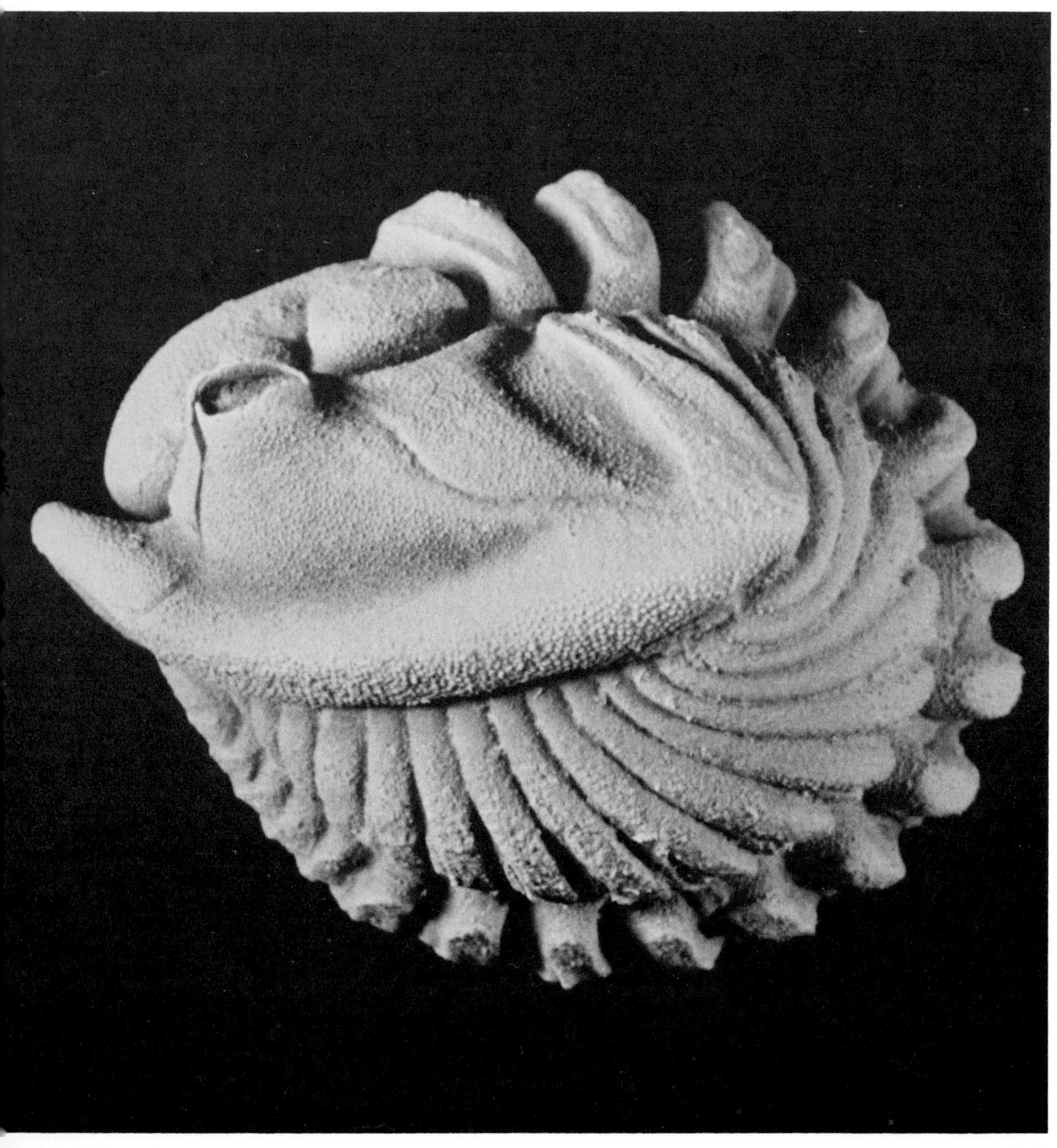

Pl. 35

better than this in rolling up. The seal was very tight, however, as can be seen in plate 33, in spite of the limitations.

An example of uncoiled spiral enrollment is shown in plate 35. The trilobite is *Flexicalymene meeki* (Foerste), Ordovician, from Waynesville, Ohio. Although the enrollment resembles the spheroidal type, it has been shown by Bergström that this particular form is in fact at the end of the line of the evolution of spiral enrollment. The tip of the pygidium is safely interlocked in a groove beneath the frontal cephalic doublure. The lateral cephalic margin contributes to wrap the pleural basket. The expansion of the axial rings and exposure of the articulating half rings is clearly visible in this beautifully preserved trilobite. It is the author's experience that enrollment helped the trilobite survive even as a fossil. The little ball was found intact after having been washed away from its burial sediment and down a steep ravine by a trickle of water. The other trilobites which were not tightly enrolled (and which were not molts) would irreparably disintegrate in the same exposure.

3.5. Life Habits

An accurate reconstruction of the mode of life of trilobites cannot, of course, be given, since we only know them as fossils. Putting together bits and pieces of evidence however, as in a detective story, we can arrive at a fairly plausible picture of trilobite habits. First of all, we know that trilobites were exclusively marine animals. Their habitat must have varied over a wide range of conditions, as can be conjectured on the basis of their adaptation. The morphology of the trilobite itself is one of the primary criteria in deducing life habits. Further

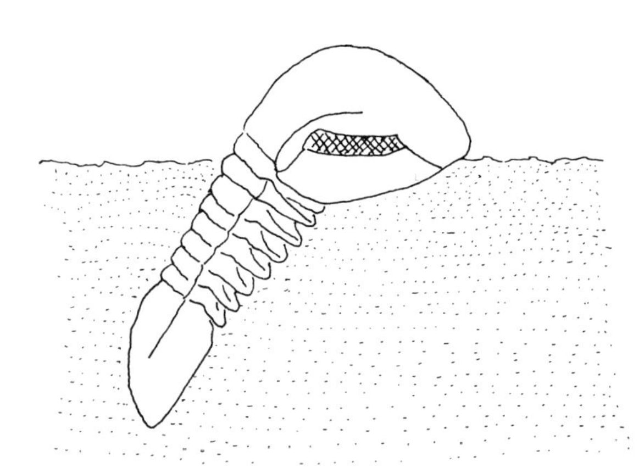

Fig. 13

Fig. 13. Probable life posture of the illaenid trilobite *Panderia megalophthalma* Linnarson. Adapted from Bergström 1973a.

evidence at our disposal is represented by faunistic associations, type of bed sediments in which the fossil was located, and ultimately the trilobite's own footprints. The description of the ventral appendage apparatus in *Triarthrus* was discussed in section 3.2 in connection with feeding habits. This is an example of the kind of inference which can be made in a particular case. Although this trilobite did not qualify as a predator but as a rather timid small-particle feeder, there are other trilobites provided with better jaws, and there is evidence that some trilobites actually hunted down their prey in soft-bottom sediments or in the water. In general, trilobites crawled on the sea floor, leaving well-known trail patterns. In a few cases the fossil trail leads to the trilobite itself, nested in its own burrow. Because of their many telepods, actual locomotion for trilobites was a by-product of sifting and reworking the soft substratum in search for food. Some pulled their bodies sideways in this process; others would use their telepods to burrow more deeply to reach for located prey. In other instances, burrows indicate a good resting or hiding place, or an observation post from which to wait for approaching prey. Burrowing became a mode of life for certain groups of trilobites, which developed a morphology particularly adapted to this purpose, such as a smooth exterior and broad axial lobe. The smooth exterior has the clear implication of reducing friction, and the wide rachis must have housed powerful appendage muscles essential to efficient burrowing. Some illaenid and asaphid trilobites have been found in what is thought to correspond to their life posture. The cephalon would rest on the surface, while the rest of the body projected downwards, as shown in figure 13 following the description by Bergström (1973a).

Active swimming on the other hand must have been the main occupation of other groups of trilobites. In these, the enrollment capacity is reduced, the body is slender and lighter, the pygidium small, the cephalon built hydrodynamically to favor laminar flow. In general the swimming trilobites had large eyes, with a field of view spanning a circular horizon. P. E. Raymond (1939) must have thought of these characters when he referred to trilobites as the butterflies of the seas. The extreme extension of the visual surface to the ventral side, as seen in section 3.3 for *Pricyclopyge* and others, is telltale evidence of swimming life. If we should learn anything from the horseshoe crab, we might infer that some trilobites liked to swim upside-down. With their carapace functioning as a glider in this attitude, their dorsal eyes could scan the sea floor for prey more efficiently than in the upright position. This, however, is only speculation, after having seen the horseshoe crab, particularly the young, frolic and swim upside-down in a marine tank at Hinds Laboratory for the Geophysical Sciences at the University of Chicago. One can also speculate that the turret like eyes of many trilobites, protruding as they did from the cephalic surface, could have been very profitably used as watchtowers above the sea floor level and still enable complete concealment of the hunter below a layer of sand. The appearance of a hard carapace itself, the enrollment capabilities, and various forms of spinosity are responses to a need for protection against an environment which may have become hostile in many ways. On the whole we must conclude that 300 million (and uncounted more) years of survival are good evidence of successful trilobite adaptation.

Fig. 14. Outline of trilobite classification following Bergström 1973a. Radiating lines indicate the geologic occurrence of the various groups. Trilobite drawings refer to characteristic representatives of each major group. Clockwise, these are *Olenellus, Paradoxides, Zacanthoides, Proetus, Zeliszkella, Arctinurus, Ceraurus, Flexicalymene, Peronopsis.*

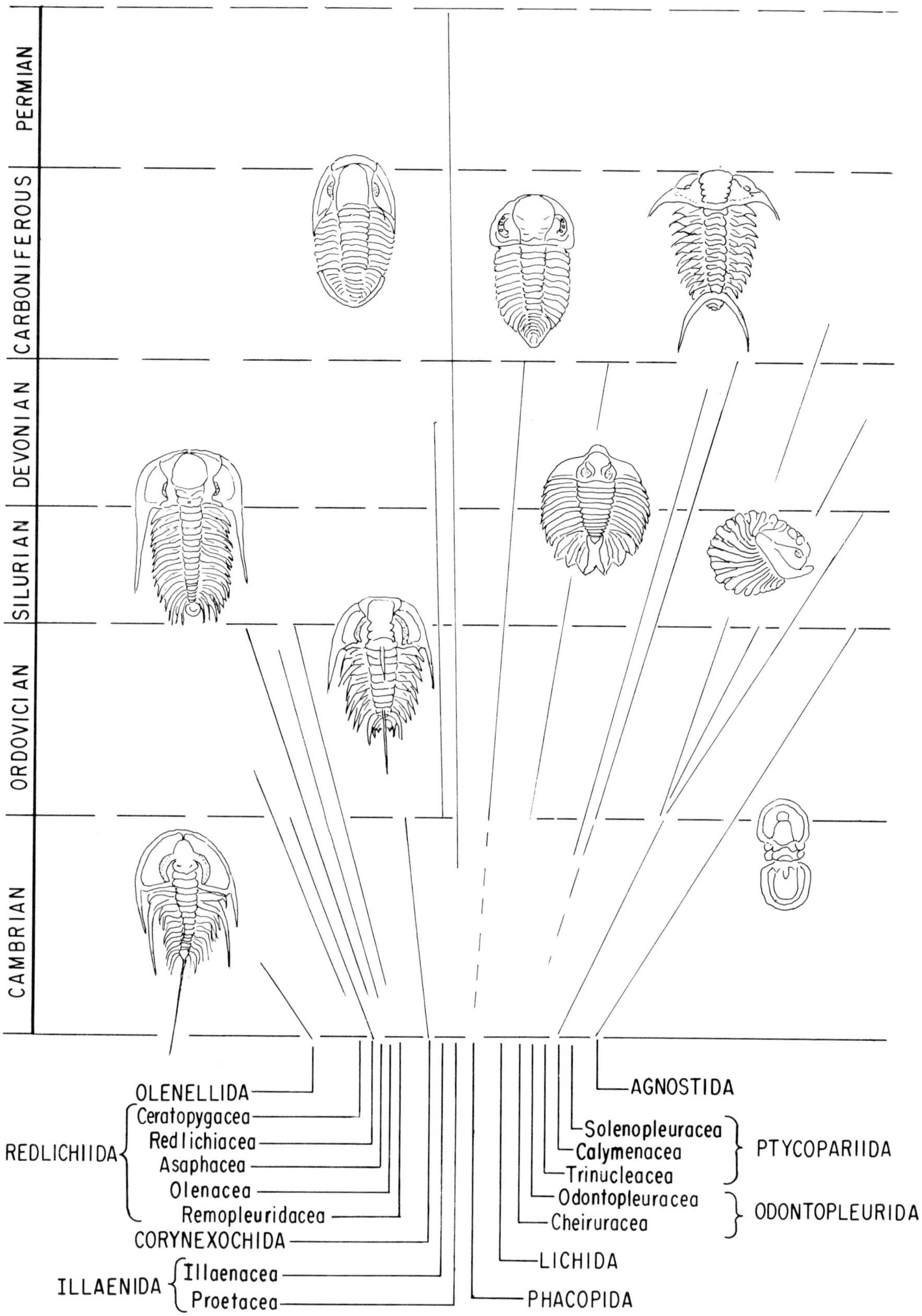

PERMIAN
CARBONIFEROUS
DEVONIAN
SILURIAN
ORDOVICIAN
CAMBRIAN

OLENELLIDA
REDLICHIIDA
Ceratopygacea
Redlichiacea
Asaphacea
Olenacea
Remopleuridacea
CORYNEXOCHIDA
ILLAENIDA
Illaenacea
Proetacea
PHACOPIDA
LICHIDA
ODONTOPLEURIDA
Cheiruracea
Odontopleuracea
PTYCOPARIIDA
Trinucleacea
Calymenacea
Solenopleuracea
AGNOSTIDA

Fig. 14

3.6. Trilobite Classification

The systematization of the fossil record of trilobites into a scheme that reflects the phylogeny and evolution of different stocks in a consistent manner has been the constant preoccupation of paleontologists since the early nineteenth century. Even the most basic subdivision of the material into major groups has been the subject of great controversy. The problem of properly assigning the parenthood and relationships of the 1500 or more genera known today, encompassing some 10,000 different species, is a monumental task indeed.

The discovery of particular anatomical or functional features in trilobites has led, in the past, to attempts at classification based on such features. The presence or absence of eyes (Dalman 1827), the eye structure (Emmrich 1839), and the enrollment ability (Milne Edwards 1840) are only a few examples of the criteria which played a role in the earliest attempts of classification. The list is long indeed and is discussed in detail in the *Treatise* (Moore 1959). In the search for a "natural classification," Beecher (1897) believed that the natural sequence of evolutionary events could be unraveled by studying the ontogenetic development of the species (Haeckel's law of morphogenesis). This led Beecher to subdivide the trilobites into three main groups on the basis of the pattern of their cephalic sutures. The Proparia and Opisthoparia are two of them, already mentioned in section 3.1; in addition he included a third group, the Hypoparia, which have ventral cephalic sutures. The first two groups have formed the basis of numerous subsequent classifications, though not without objection from other paleontologists. The third group has been rejected altogether. In the more recent attempts at classification, it has been recognized that no individual feature, no matter how significant it may be, is sufficient guide for classification, and that affinities based on collective characteristics must be weighed. The cephalic sutures have always been considered a significant guideline, up to the classification adopted in the *Treatise*, together with the cephalic axial characters. Further discussion on this line, however, threatens to become too technical and beyond the scope of this work. For this we refer the reader to the *Treatise* (Moore 1959) and to some of the literature which will be referred to in the following discussion.

At the Oslo Trilobite Conference of July 1973 I met Dr. Jan Bergström of the Department of Historical Geology and Paleontology, University of Lund, Sweden. In a recent comprehensive survey and innovative study (Bergström 1973a), Dr. Bergström had just reexamined critically the problem of trilobite classification. In Bergström's work the role of trilobite enrollment capability as an additional phylogenetical index has been emphasized and integrated with previously used criteria. (Richter 1933; Henningsmoen 1951; Hupé 1953, 1955). As a result, a more consistent picture has emerged—one which consolidates much of the previous proliferation of orders and suborders, often based on characters of no phylogenetic significance. In many respects, Bergström's classification departs radically from the *Treatise*.

Having been brought up to date by Dr. Bergström, and still preparing this book at that time, I was tempted by the prospect of organizing the Atlas material according to the newly proposed classification. This organization was made easier by further correspondence with Dr. Bergström, who clarified for me many controversial points. Nine orders of trilobites are recognized in Bergström's work. A pictorial illustration of these groupings is given in figure 14. Here each branch describes the span of geological time encompassed from first appearance to extinction, and a sketch indicates a relevant trilobite for that particular group. The Olenellida and Redlichiida appear to have a common origin in the Early Cambrian or Late Precambrian. All other trilobites seem to be related off-springs of the redliichids.

Atlas of Trilobite Photographs 4

Following the classification scheme illustrated in section 3.6, the Atlas is organized in nine sections, each corresponding to one of the major groups (taxonomic order) recognized by Bergström (1973a). Each section is preceded by a brief description of the trilobite group according to Bergström; this description must necessarily be somewhat technical and contains a sequence of representative trilobite photographs. The captions accompanying the plates contain detailed taxonomic and stratigraphic information but also stress general and often nontechnical features.

It should be emphasized that completeness has always been remote from my intentions in the selection of the material to be included in each section. In fact, I have been very partial in this choice, presenting a large number of specimens from my own personal collection. Although a significant number of trilobites have been borrowed from various museum collections, to extend the range of coverage, the Atlas is still extremely lopsided, with obvious emphasis on available material. In no way is the Atlas intended as an encyclopedic description of the subject. Furthermore, I could not refrain from occasionally presenting more than one example of the same trilobite in order better to convey a feeling for the range of occurrences and to establish contact with the fossil collector as well as with the paleontologist. At times photographs have been chosen simply for their aesthetic appeal to me, and I have also taken irreverent liberties in choosing only perfect specimens for presentation and in deliberately using nonconventional photographic techniques.

The main preoccupation has been that of presenting more immediate evidence of trilobites than, for example, that provided by the accurate drawings which adorn the professional literature. This may help give a measure of hope to the fossil collector who has just unearthed a precious but frustrating trilobite fragment. Another preoccupation has been that of the photographic format. Trilobites are generally small, and reproductions in natural size tend to become insignificant. Adequate enlargement often reveals a lot more than the unaided eye can see. In conclusion, I wish to point out with these pictures that there were other exciting prehistoric animals besides dinosaurs. Trilobites may deserve some excitement on our part as well.

Explanatory Note on Captions to Atlas Plates

Each plate is assigned coded initials indicating order, geologic column assignment, and progressive numeral, according to the abbreviation code tabulated on page 00. The tabulation also includes the key to the word terminations which are used to denote taxonomic classification in invertebrate paleontology. In each legend the generic attribution (genus) is indicated first, followed by synonyms, if any, or other specifications. The combination of generic and trivial name for the species represented in the plate is indicated in *italics*. Whenever more than one species belonging to the same genus are presented in sequence, only the legend of the first plate of the sequence carries the generic name, author, and date. With the exceptions indicated in the captions, all photographs have been prepared by the author. Photographic techniques and specimen preparation are described in the Appendix B. Unless a special technique is mentioned, the specimens have been photographed in air and uncoated. The paleontological convention of accentuating the illumination of specimens with light coming from the northwest has not been followed whenever a more profitable rendition of details could otherwise be obtained.

Many of the specimens originate from the University of Chicago Walker Museum, now transferred to the Field Museum of Natural History in Chicago. The disposition of this material and others will be indicated in abbreviated form (see code). The responsibility for the classification and identification of specimens indicated by the term *RLS id.* rests solely with the author.

Abbreviation code

Order	*Code*
OLENELLIDA	OL
PTYCHOPARIIDA	PT
REDLICHIIDA	RE
ILLAENIDA	IL
PHACOPIDA	PH
ODONTOPLEURIDA	OD
LICHIDA	LI
CORYNEXOCHIDA	CO
AGNOSTIDA	AG
Geologic Column	
CAMBRIAN	CAM
ORDOVICIAN	ORD
SILURIAN	SIL
DEVONIAN	DEV
CARBONIFEROUS	CARB
PERMIAN	PERM
Miscellaneous	
Identified	id.
Collection	coll.
University of Chicago Walker Museum	UCWM
Museum of Comparative Zoology Harvard University	MCZ
Peabody Museum, Yale University	YPM
Field Museum of Natural History, Chicago	FMNH

Taxonomic Nomenclature

Taxa	*Termination*	*Example*
Class	a	Trilobita
Order	ida	Phacopida
Suborder	ina	Phacopina
Superfamily	acea	Phacopacea
Family	idae	Phacopidae
Subfamily	inae	Phacopinae
Genus	. . . .	*Phacops*
Species	. . . .	*rana*
Variety	. . . .	*milleri*

Class Trilobita Walch, 1771

4.1. Order Olenellida Resser, 1938 (emended Bergström 1973a)

Emended diagnosis (Bergström 1973b): Trilobites with perrostral suture and sickle-shaped rostral plate; hypostome attachment variable; compound eyes long, generally extending along the entire length of the palpebro-ocular ridges; median eyes not observed; palpebro-ocular ridge generally connected with glabella; palpebral area commonly narrow, particularly in convex forms; glabellar furrows distinct. Thorax tapering, in several forms divisible into prothorax and narrow opisthothorax; pleurae sloping distal to dorsal furrow or line, without long horizontal hinge. Pygidium generally small, commonly formed by postsegmental telson and possibly a few segments. Enrollment capacity lacking in some forms, poorly developed in others.

The classification proposed by Bergström includes the following taxa:

- Superfamily Olenellacea Vogdes, 1893
  - Daguinaspididae Hupé, 1953
    - Daguinaspidinae Hupé, 1953
    - Fallotaspidinae Hupé, 1953
    - Nevadiinae Hupé, 1953
    - Neltneriinae Hupé, 1953
    - Callaviinae Poulsen, 1959
  - Holmiidae Hupé, 1953
  - Olenellidae Vogdes, 1893

The *emended diagnosis* for the Olenellidae, the main group of trilobites represented here, is as follows: Olenellacean trilobites with fairly large frontal glabellar lobe; palpebral area narrow; metagenal points distally positioned; pleural furrows generally extending through pleural spines; third thoracic tergite may be macropleural; hypostome commonly multidentate. Not enrolling.

Order: **Olenellida Resser, 1938 (emended Bergström, 1973a)**

Superfamily: **Olenellacea Vogdes, 1893**

Family: **Olenellidae Vogdes, 1893**

Plate 36. OLCAM 1. OLENELLUS Billings, 1861 (= *Fremontia* Raw, 1936; *Mesonacis* Walcott, 1885; *Paedumias* Walcott, 1910), *Olenellus fremonti* Walcott (x8). Lower Cambrian of British Columbia. *Bonnia-Olenellus* assemblage zone in the St. Piran sandstone (Peyto limestone member). (Gift of J. R. Evans to UCWM; loaned by FMNH; id. RLS.) The classification of the Olenellidae has undergone extensive revision in recent years (Fritz 1972; Bergström 1973b). Having changed generic affiliation several times, *Olenellus fremonti* is presently reinstated with the original denomination given by Walcott, 1910. Note exaggerated development of the pleural lobe of the third thoracic segment. The irregular fracture along the axis is probably due to compression. The axial spine is hollow.

Pl. 36

Plate 37. OLCAM 2. Another specimen of *Olenellus fremonti* Walcott (x5.7) from the same sample which yielded OLCAM 1. The carapace is incomplete and flattened by compression. As in OLCAM 1, the entire exoskeleton is distorted by shear stress in the sediment.

Pl. 37

Plate 38. OLCAM 3. Juvenile form of *Olenellus fremonti* Walcott (x11), again from the sample which yielded OLCAM 1 and 2. (RLS coll., id.; courtesy FMNH.) In (a) the specimen is photographed by standard technique, while in (b) the specimen is photographed while immersed in xylene. The optical contact of this medium (index of refraction n = 1.5) with the surface crystalline layers of the sample provides a much better representation of the anatomical features than in (a). Note that the axial spine is as long as the entire carapace.

Pl. 38a

Pl. 38b

Plate 39. OLCAM 4a. *Olenellus clarki* (Resser) (x7). Lower Cambrian of Pioche, Nevada. Pioche shale, D. Member. Specimen collected by A. Fawcett, during a field trip with the author. (RLS coll., id.) Part of a slab containing four partially complete individuals. Specimen whitened with magnesium oxide. This trilobite was previously assigned to the Genus *Paedumias* Walcott, 1910 (see synonyms mentioned for OLCAM 1) and bears great resemblance to *Olenellus transitans* (Walcott) of the Atlantic faunal province. Note wide frontal area anterior to glabella, marked by median ridge, and third segment macropleurae. The latter, however, are not as wide as in *O. fremonti* (Walcott).

Pl. 39

Pl. 40

Plate 40. OLCAM 4b. The negative cast, counterpart of the internal mold of OLCAM 4a. A common optical illusion (as in photographs of the Moon's craters) can make the concave surface appear convex.

***Subfamily:* Wanneriinae Hupé, 1953**

Plate 41. OLCAM 5. WANNERIA Walcott, 1910 *Wanneria* sp. (x2). Lower Cambrian, Silver Peak, Nevada. (Loaned by MCZ. RLS id.) This specimen, originally attributed to *Olenellus* sp., seems to fit the description of *Wanneria* much better. Note reticulated surface on pleurae and cephalon, terrace lines on pleural spines. The cephalon is too compressed and distorted to enable specific classification.

Pl. 41

4.2. Order Redlichiida Richter, 1933 (emended Bergström, 1973a)

Emended diagnosis: With few exceptions, trilobites with opisthoparian sutures and medium-sized nonswollen glabella; eyes large in early forms but tend to be smaller in many later members; hypostome commonly connected with rostral plate, rarely fused; rostral plate tends to disappear; thorax commonly with pointed pleural spines; horizontal hinge-line present in early forms but disappears in some instances; flanges almost invariably absent; ring joint and dorsal furrow joint common; pygidium small to large, commonly with flattened border and marginal spines; enrolling capacity lacking in forms without hinge-line, spheroidal or cylindrical in others; panderian organ or vincular apparatus may limit enrolling action, in which pleural spines slide over one another.

Taxa included are:

Suborder REDLICHIIDA Richter, 1933.

—Redlichiid trilobites mostly without panderian organ and vincular apparatus; where medium-sized or large, the pygidium more commonly than not has marginal spines.

Superfamily Redlichiacea Poulsen, 1927 (emended Bergström, 1973a).

—Redlichiine trilobites commonly with tapering glabella having simple parallel glabellar furrows; palpebro-ocular ridge simple except in some early members, either fairly transverse or inclined strongly backwards; posterior branch of facial suture short and longitudinal to long and transverse; pygidium commonly small.

Protolenidae Richter & Richter, 1948
- Bigotininae Hupé, 1953
- Termierellinae Hupé, 1953
- Protoleninae Richter & Richter, 1948
- Myopsoleninae Hupé, 1953
- Palaeoleninae Hupé, 1953
- Lermontoviinae Suvorova, 1956
- Bergeroniellinae Repina, 1966

Aldonaiidae Hupé, 1953
Metadoxididae Whitehouse, 1939
Jakutidae Suvorova, 1958
Redlichiidae Poulsen, 1927
- Neoredlichiinae Hupé, 1953
- Abadiellinae Hupé, 1953
- Wutingaspidinae Chang, 1966
- Pararedlichiinae Hupé, 1953
- Redlichiinae Poulsen, 1927

Despujolsiidae Harrington, 1959 (= Resseropidae, Chang, 1966)
- Despujolsiinae Harrington, 1959
- Resseropinae Chang, 1966

Dolerolenidae Kobayashi, 1951
Yinitidae Hupé, 1953 (= Drepanopygidae Lu, 1961)
- Yinitinae Hupé, 1953
- Drepanopyginae Lu, 1961

Mayiellidae Chang, 1966
Gigantopygidae Harrington, 1959
Emuellidae Pocock, 1970
Paradoxididae Hawle & Corda, 1847
- Paradoxidinae Hawle & Corda, 1847
- Xystridurinae Whitehouse, 1847
- Centropleurinae Angelin, 1854

Bathynotidae Hupé, 1953
Burlingiidae Walcott, 1908

Superfamily Ceratopygacea Linnarsson, 1869 (emended Bergström, 1973).

—Redlichiine trilobites generally with tapering glabella; glabellar furrows tend to divide or get complicated in other ways; plectrum commonly present; posterior branch of facial suture long, transverse; hingeline present; pygidium medium-sized to large, commonly with marginal spines; enrollment spheroidal, if present.

Papyriaspididae Whitehouse, 1939
Ceratopygidae Linnarsson, 1869
Mapaniidae Chang, 1963
Asaphiscidae Raymond, 1924
- Asaphiscinae Raymond, 1924
- Blountiinae Lochman, 1944

Tsinaniidae, Kobayashi, 1933
Erixaniidae Öpik, 1963
Plectriferidae Öpik, 1967
Rhyssometopidae Öpik, 1967
Chanshaniidae Kobayashi, 1935
Damesellidae Kobayashi, 1935
Polycyrtaspididae Öpik, 1967
Auritamidae Öpik, 1967
Liostracinidae Raymond, 1937
- Liostracininae Raymond, 1937
- Doremataspidinae Öpik, 1967

Harpididae Whittington, 1950
Kaolishaniidae Kobayashi, 1935
- Kaolishaniinae Kobayashi, 1935
- Manuyiinae Hupé, 1955
- Tingocephalinae Hupé, 1955

Marjumiidae Kobayashi, 1935
Crepicephalidae Kobayashi, 1935
Tricrepicephalidae Palmer, 1954

Suborder ASAPHINA Salter, 1864 (emended Bergström, 1973).

—Redlichiid trilobites commonly with long posterior branch of facial suture; rostral plate commonly disappeared, doublure repeatedly fused over median suture; well-developed hingeline if not secondarily lost; commonly with panderian organ; pygidium generally medium-sized to large, with or without marginal spines; if spineless, commonly fitting cephalon margin to margin in enrollment and of identical outline; enrollment spheroidal, secondarily cylindrical, or ability lost.

Superfamily Asaphacea Burmeister, 1843 (emended Bergström, 1973).

—Asaphine trilobites with a fixed hypostome, commonly with a median incision or spine in the hypostomal margin; genal spines commonly strong.

- Anomocaridae Poulsen, 1927
  - Anomocarinae Poulsen, 1927
  - ?Conokephalininae Walcott, 1913
- Andrarinidae Raymond, 1937
- Parabolinoididae Lochman, 1956
- Loganellidae Rasetti, 1959
- Idahoidae Lochman, 1956
- Dikelocephalidae Miller, 1889
  - Dikelocephalinae Miller, 1889
  - Saukiinae Ulrich & Resser, 1930
- Dikelocephalinidae Kobayashi, 1936
- Asaphidae Burmeister, 1843
  - Taihungshaniinae Sun, 1931
  - Asaphinae Burmeister, 1843
  - Isotelinae Angelin, 1854
  - Niobinae Jaanusson, 1959
  - Ogygiocaridinae Raymond, 1937
  - Promegalaspidinae Jaanusson, 1959
  - Thysanopyginae Jaanusson, 1959
  - Griphasaphinae Öpik, 1967
  - ?Symphysurininae Kobayashi, 1935
- Nileidae Angelin, 1954

Superfamily Olenacea Burmeister, 1843.

—Asaphine trilobites with hypostome probably disconnected from doublure; genal spines needlelike.

- Olenidae Burmeister, 1843
  - Oleninae Burmeister, 1843
  - Leptoplastinae Angelin, 1854
  - Pelturinae Hawle & Corda, 1847
  - Triarthrinae Ulrich, 1930
  - Rhodonaspidinae Öpik, 1963
  - ?Talbotinellinae Öpik, 1963
- Hypermecaspididae Harrington & Leanza, 1957

Superfamily Remopleuridacea Hawle & Corda, 1847.

—Asaphine trilobites with subquadrate hypostome fixed to doublure; anterior wing process obviously extending from doublure, not from hypostome; palpebral area narrow, tends to merge with glabella; hingeline tends to disappear completely and strong fulcral process and socket pivot joint takes the position of the dorsal furrow joint; pygydium small to medium-sized, generally with marginal spines.

- Remopleurididae Hawle & Corda, 1847
  - Remopleuridinae Hawle & Corda, 1847
  - Richardsonellinae Raymond, 1924
- Hungaiidae Raymond, 1924
- Bohemillidae Barrande, 1872

Order: **Redlichiida Richter, 1933 (emended Bergström, 1973a)**

Suborder: **Redlichiina Richter, 1933**

Superfamily: **Redlichiacea Poulsen, 1927**

Family: **Dolerolenidae Kobayashi, 1951**

Plate 42. RECAM 1. DOLEROLENUS Leanza, 1949. *Dolerolenus zoppii* (Meneghini) (x10) Up. L. Cambrian, Porto Canalgrande, Sardinia, Italy. (RLS coll., id.) Specimen whitened with magnesium oxide.

Pl. 42

Plate 43. RECAM 2. *Dolerolenus zoppii* (Meneghini) (x2.6), from the same locality as RECAM 1. (RLS coll., id.)

Pl. 43

Family: **Paradoxididae Hawle and Corda, 1847**

Plate 44. RECAM 3a. PARA-DOXIDES Brongniart, 1822. *Paradoxides gracilis* (Boeck) (x2.5). M. Cambrian, Jinetz, Bohemia. (RLS coll.; gift of Mr. M. Gazay).

Pl. 44

Plate 45. RECAM 3b. Enlarged detail of the slab figured in the preceding plate (x3.6).

Pl. 45

Pl. 46

Plate 46. RECAM 4. *Paradoxides gracilis* (Boeck) (x5). M. Cambrian, Jinetz, Bohemia. (RLS coll.; courtesy of MCZ.) Specimen whitened with magnesium oxide.

Superfamily: **Ceratopygacea Linnarson, 1969 (emended Bergström, 1973a)**

Family: **Papyriaspididae Whitehouse, 1939**

Plate 47. RECAM 5. ALOKISTOCARE Lorenz, 1906. *Alokistocare harrisi* Robison (x7.2). Wheeler shale formation, M. Cambrian, Antelope Springs, Utah. (Collected by A. Fawcett. RLS coll., id.) Following Öpik (1961), the family Alokistocaridae Resser, 1939, is regarded as a younger synonym of Papyriaspididae Whitehouse, 1939. Specimen whitened with magnesium oxide.

Pl. 47

Plate 48. RECAM 6. *Alokistocare piochensis* (Walcott) (x4.4). Chisholm shale, M. Cambrian, Half Moon Mine at Pioche, Nevada. (Collected by A. Fawcett. RLS coll., id.) Specimen photographed while immersed in xylene. One of the free cheeks is displaced.

Pl. 48

Plate 49. RECAM 7. *Alokistocare idahoensis* Resser (x4.4). Spence shale, M. Cambrian, Liberty, Idaho. (RLS coll.) External impression. Photograph printed from a color slide, to take advantage of color contrast in the specimen.

Pl. 49

Plate 50. RECAM 8. ELRATHIA Walcott, 1924. *Elrathia kingii* (Meek) (x2). Wheeler shale formation, Antelope Springs, Millard Co., Utah. (Collected by A. Fawcett. RLS coll.) In this fossilization the trilobite exoskeleton is replaced by calcite; the crystals of this mineral extend in a so-called "cone-in-cone" structure well beneath the original carapace.

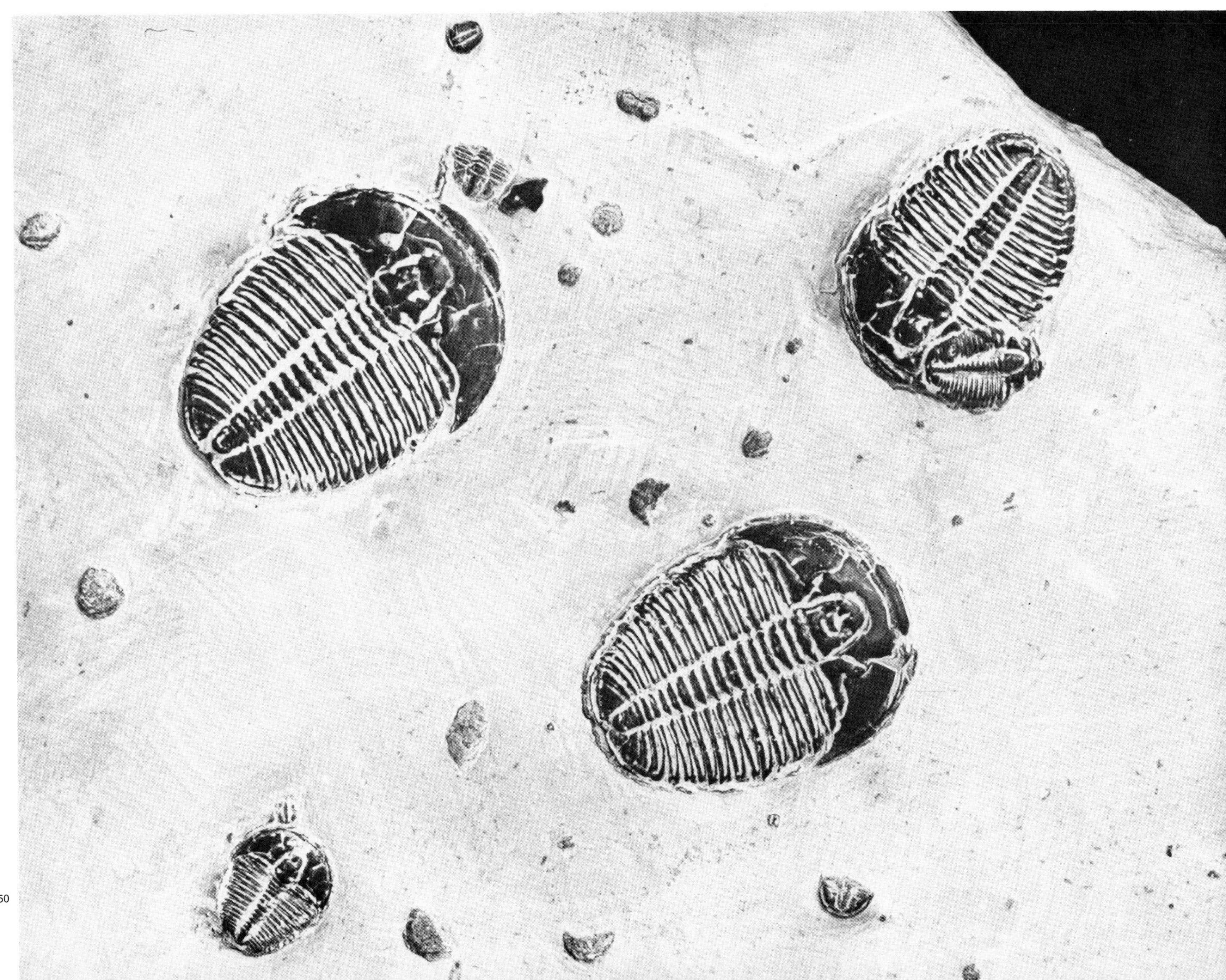

Plate 51. RECAM 9. *Elrathia kingii* (Meek) (x3.9). Same origin as for RECAM 8. (Collected by A. Fawcett. RLS coll.) Specimen of approximately maximum size for this species.

Pl. 51

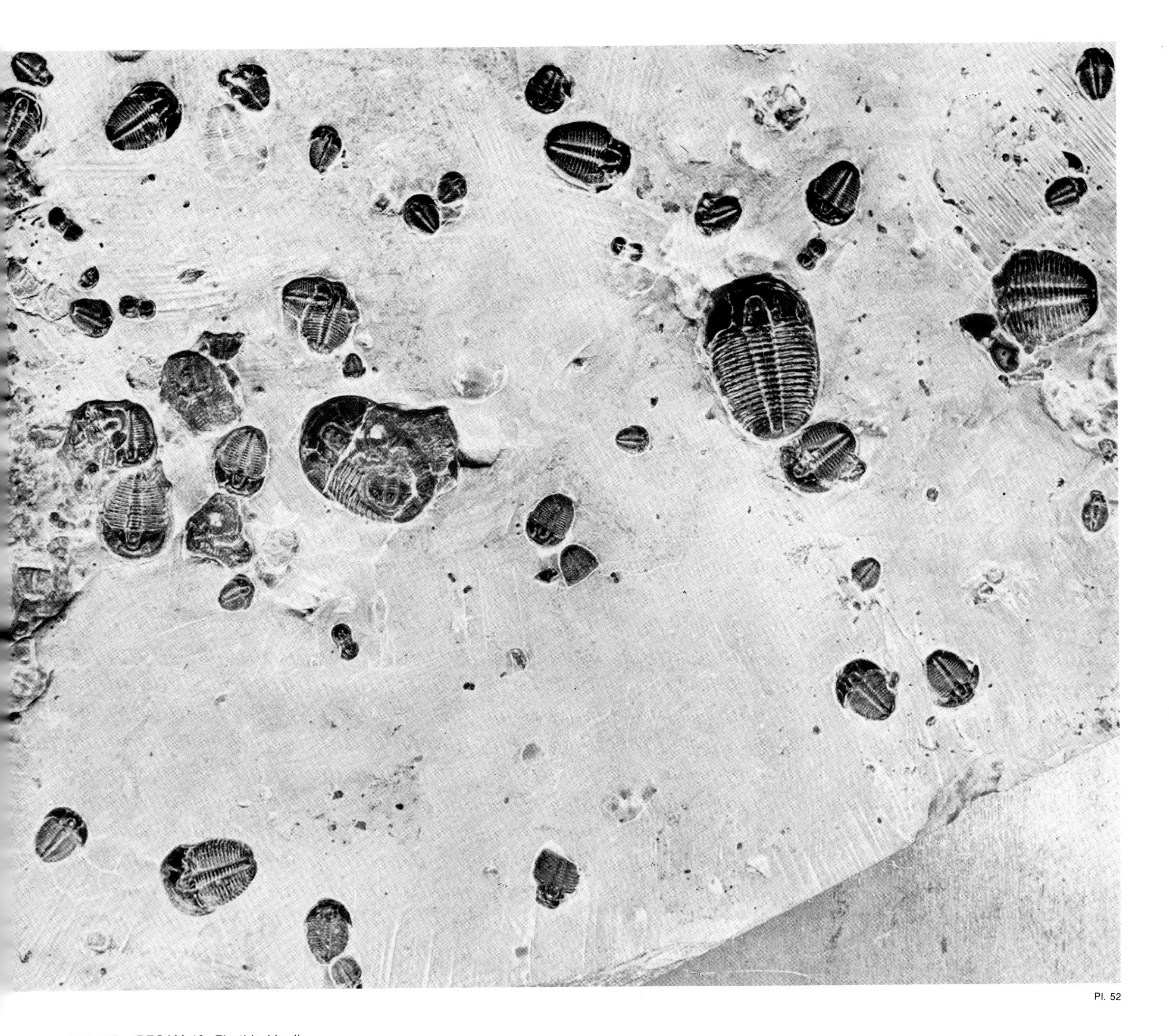

Pl. 52

Plate 52. RECAM 10. *Elrathia kingii* (Meek) in association with the agnostid trilobite *Peronopsis interstricta* (White) (x.9). (Collected by A. Fawcett. RLS coll.) Various stages of growth of *Elrathia* are represented in this unusual plate.

Family: **Asaphiscidae Raymond, 1924**

Subfamily: **Asaphiscinae Raymond, 1924**

Plate 53. RECAM 11. ASAPHISCUS (Meek) 1873, *Asaphiscus wheeleri* Meek (x3.3). Wheeler Formation, M. Cambrian, Antelope Springs, Millard Co., Utah. (Collected by A. Fawcett. RLS coll.) In this group, *Asaphiscus* is associated with *Elrathia kingii*.

Pl. 53

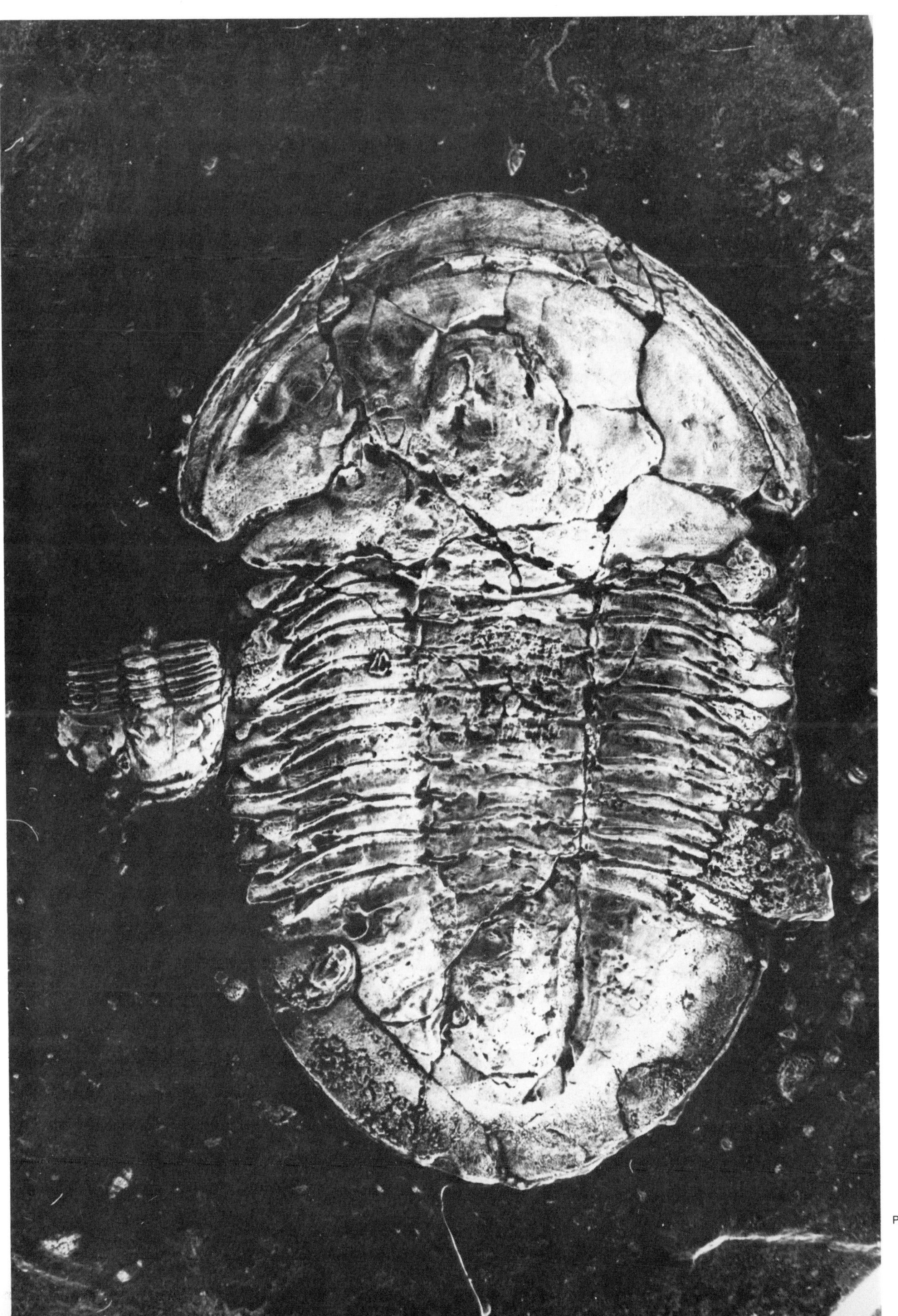

Plate 54. RECAM 12. *Asaphiscus wheeleri* Meek (xl.9). Same origin as for RECAM 11. (RLS coll.) This is an extraordinarily large specimen. Photograph printed from color slide. The original specimen appears black, as for RECAM 11.

Pl. 54

Suborder: **Asaphina Salter, 1864 (emended Bergström, 1973a)**

Superfamily: **Asaphacea Burmeister, 1843 (emended Bergström, 1973a)**

Family: **Asaphidae Burmeister, 1843**

Subfamily: **Isotelinae Angelin, 1854**

Plate 55. REORD 13. ISOTELUS De Kay, 1824. *Isotelus gigas* De Kay (x3.2). Trenton group, Ordovician, Trenton Falls, N.Y. (UCWM, loaned by FMNH.) Specimen whitened with magnesium oxide. Since the original specimen is shiny black against a black matrix, the uncoated sample would be extremely difficult to portray

Pl. 55

Pl. 56

Plate 56. REORD 14. HOMOTELUS Raymond, 1925. *Homotelus bromidensis* Esker (x3.1). Bromide formation (Poolville member), Ordovician, Blackriverian, Criner Hills, Carter Co., Oklahoma. (RLS coll.)

Plate 57. REORD 15. *Homotelus bromidensis* Esker (xl.5). Same origin as for REORD 14. (Now at the Musée d'Histoire Naturelle, Geneva, Switzerland.) In this exceptional group of trilobites, two individuals are exposed from the ventral side, showing the hypostoma.

Pl. 57

***Subfamily:* Ogygiocaridinae**
Raymond, 1937

Plate 58. REORD 16. OGYGITES Tromelin and Lebesconte, 1876. *Ogygites canadensis* (Chapman) (x2.5). Collingswood formation, Ordovician, Collingswood, Ontario (RLS coll., id.) An incredibly dense assemblage of exuviae from one trilobite species. Two incomplete carapaces, cephalon missing, are visible here, together with separate pygidia, cephala and disarticulated pleural segments.

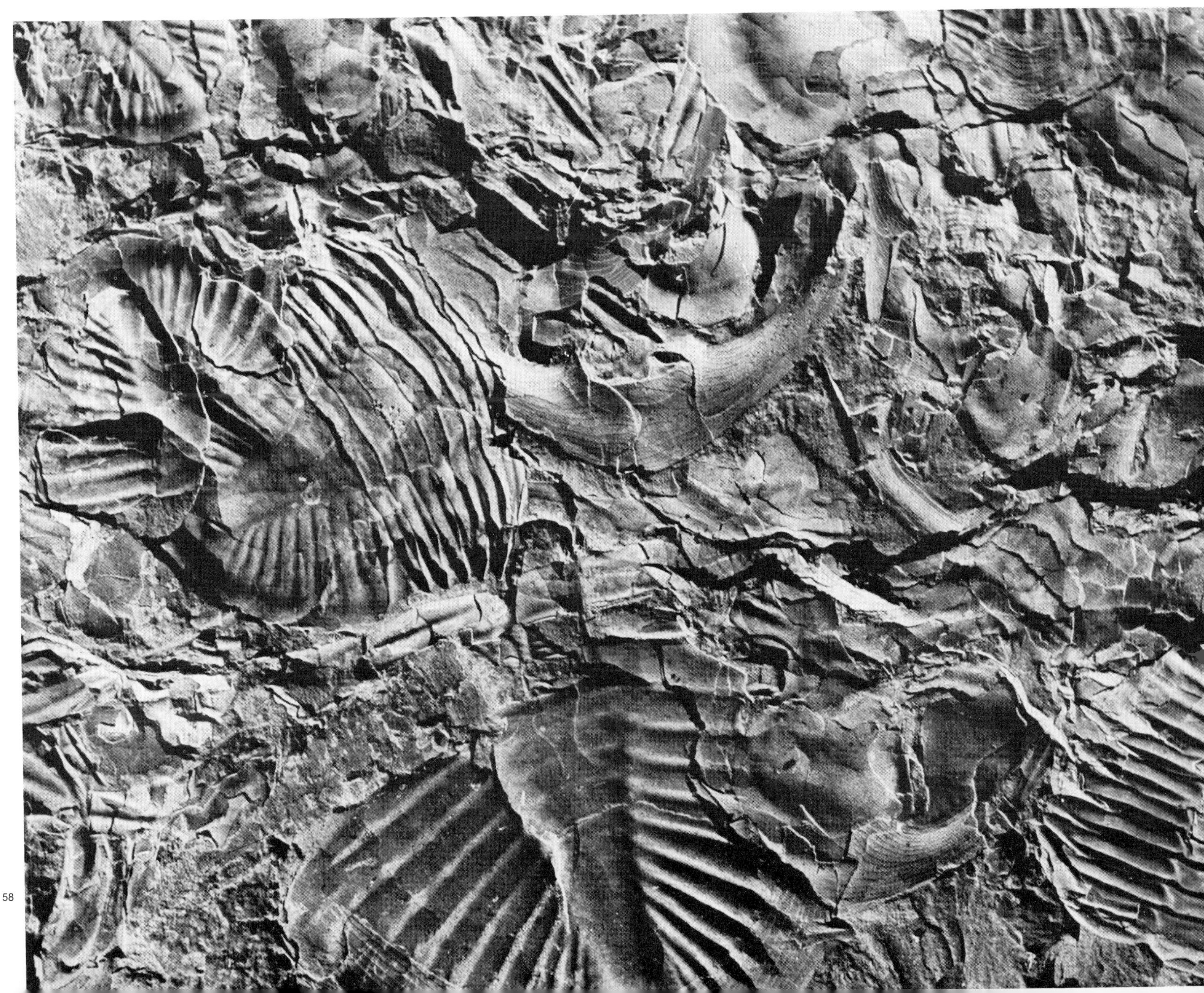

Pl. 58

Plate 59. REORD 17. Another section of the same slab yielding REORD 16 (x2.4). (RLS coll.) Several complete pygidia of *Ogygites canadensis* (Chapman) and separate cranidia can be seen.

Pl. 59

Plate 60. REORD 18. OGYGIOCARIS Angelin, 1854. *Ogygiocaris sarsi* Angelin (x3.3). Ordovician, Toten, Norway. (Original at Palaeontologisk Museum, University of Oslo.) Cast, whitened with magnesium oxide.

Pl. 60

Pl. 61

***Superfamily:* Olenacea Burmeister, 1843**

***Family:* Olenidae Burmeister, 1843**

***Subfamily:* Triarthrinae Ulrich, 1930**

Plate 61. REORD 19. TRIARTHRUS Green, *Triarthrus eatoni* (Hall) (x3). "Beecher's trilobite bed," Frankfort shale, U. Ordovician, Rome, N.Y. (YPM 218; loaned through courtesy of Dr. J. Cisne.) This and the following plates through REORD 22 represent a unique record of the Beecher's famous trilobites, which were preserved with most of the soft parts replaced by fine granules of iron pyrite. As a result such external anatomical details as antennae and appendages can be easily seen, while x-ray photographs can detect details of the internal anatomy. The photograph presented in this plate was obtained after immersion of the shale slab in xylene. The antennae and appendages become clearly discernible by this approach. One of the specimens, originally prepared by Beecher, shows the ventral side of the trilobite.

Plate 62. REORD 20. *Triarthrus eatoni* (Hall) (x7). (YPM 218E.) Same origin and disposition as REORD 19. Photograph through immersion in xylene. The biramous appendages of the right-hand side project outside the exoskeleton.

Pl. 62

Plate 63. REORD 21. *Triarthrus eatoni* (Hall) (x8.6). (YPM 204.) Same origin and disposition as REORD 19. An x-ray view (a) and a xylene immersion view (b) of the same specimen. The original radiograph was taken by Dr. J. Cisne and kindly loaned to the author. The resolution of the photograph obtained from the xylene-immersed specimen enables the fine structure in the biramous appendages, particularly the exites, to become apparent.

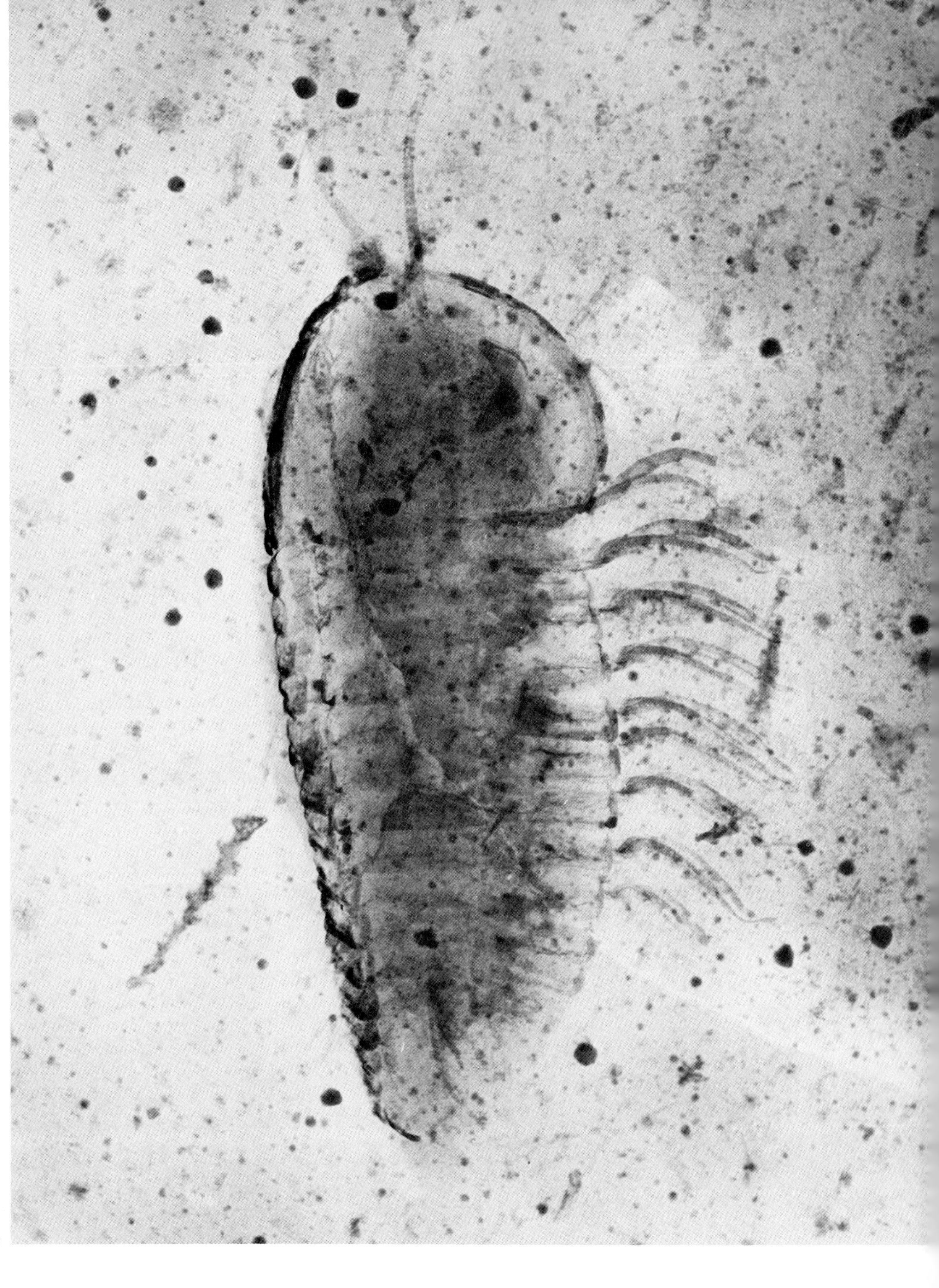

Pl. 63a

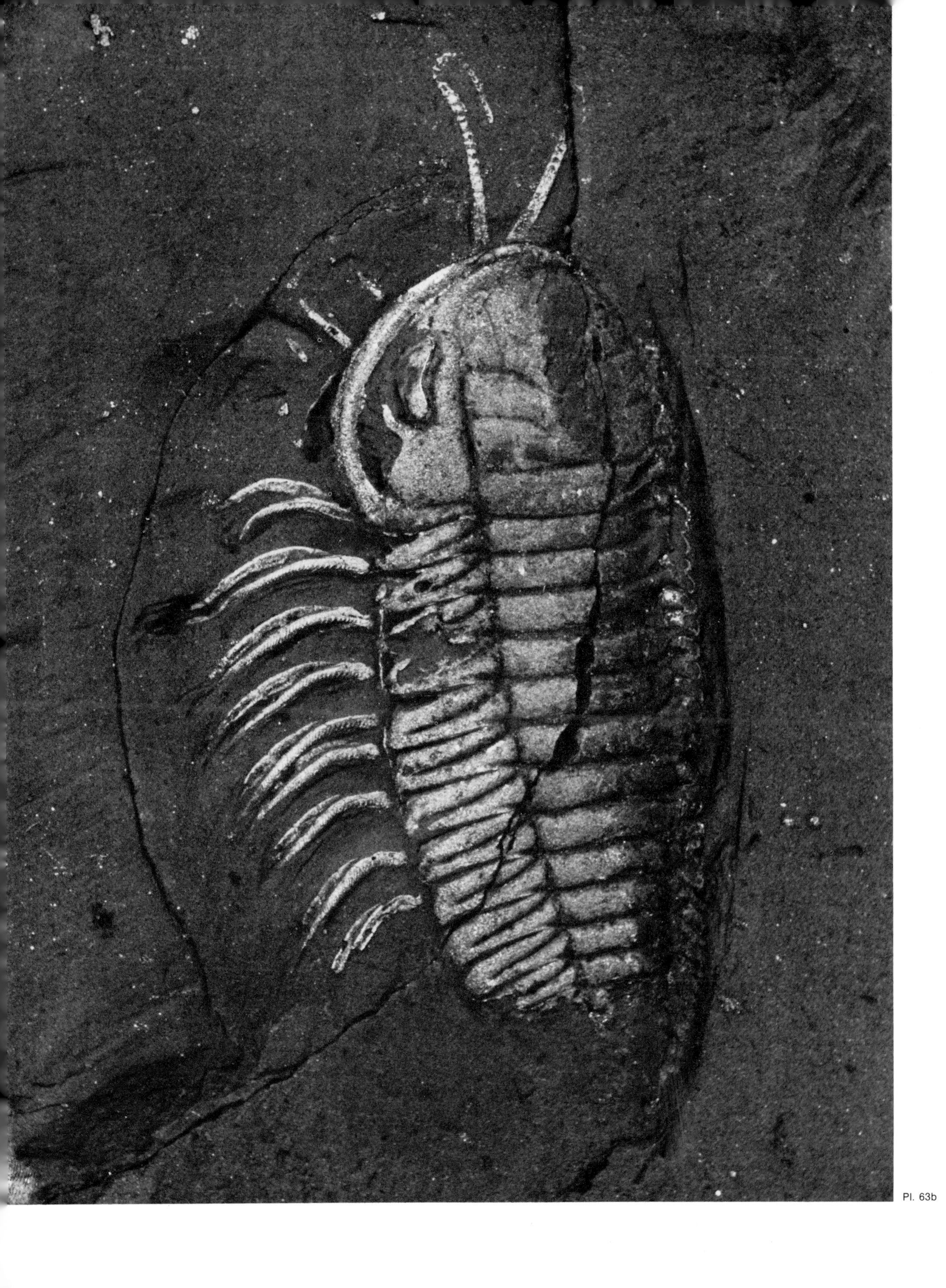

Pl. 63b

Plate 64. REORD 22. *Triarthrus eatoni* (Hall) (x4.6). Same origin as for REORD 19. (Am. Mus. Nat. Hist. 839/14 A,B through courtesy of Dr. J. Cisne. X-ray negative loaned by Dr. J. Cisne.)

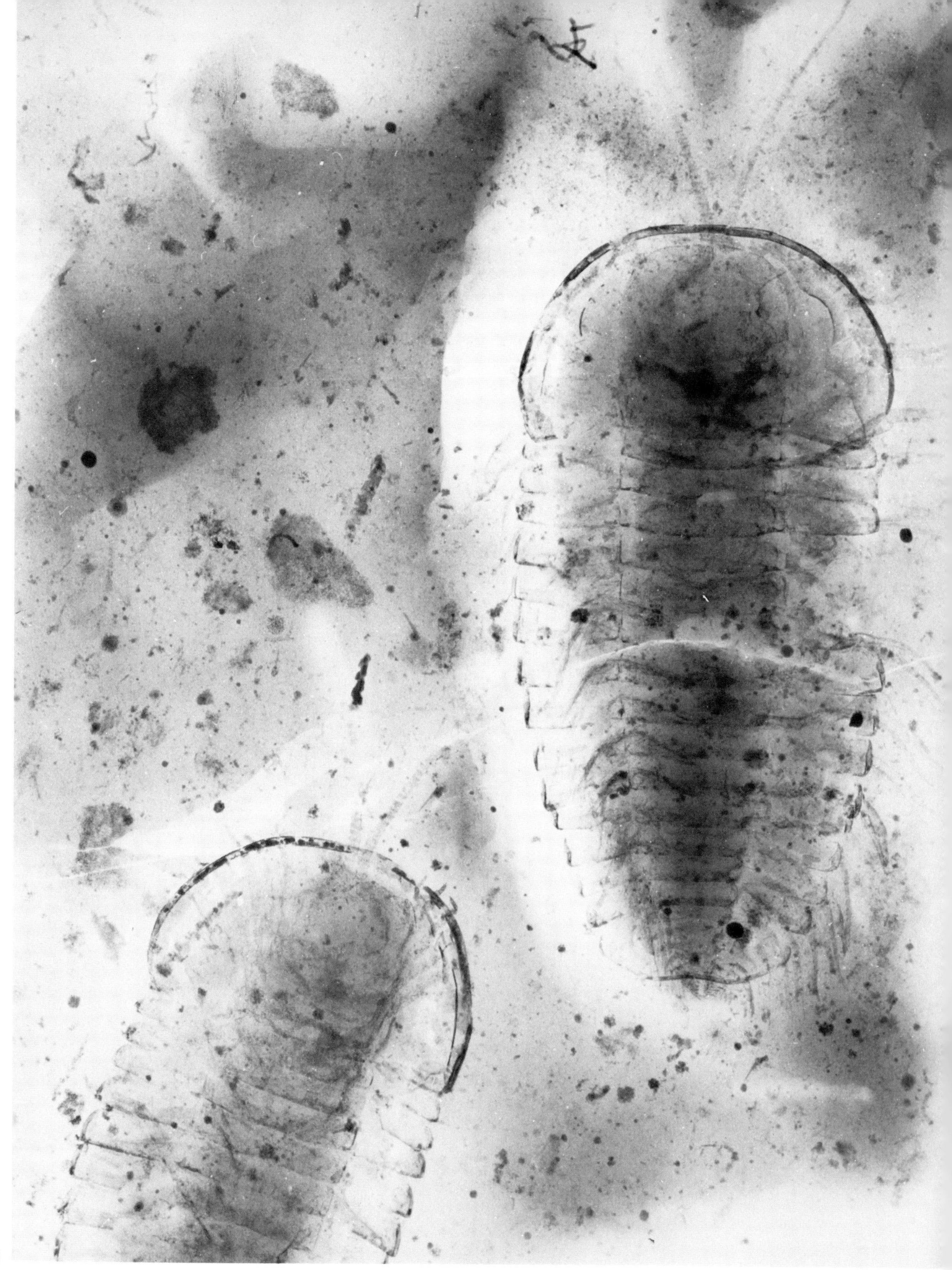

Pl. 64

4.3. Order Corynexochida Kobayashi, 1935 (emended Bergström, 1973a)

Emended diagnosis (Bergström 1973a): Trilobites with opisthoparian suture in most cases; hypostome generally fused with rostral plate; paired eyes commonly long but may be short or absent; glabella generally expanding forward or parallel-sided; glabellar furrows on each side commonly converging outward; horizontal hingeline generally developed in thorax; dorsal furrow joint and fulcral joint may be present; pygidium generally medium-sized or large, with entire or spinous border; spheroidal enrollment at least in some members.

Taxa included are:

- Superfamily Corynexochacea Angelin, 1854
  - Hicksiidae Hupé, 1953
  - Dorypygidae Kobayashi, 1935
  - Ogygopsidae Rasetti, 1951
  - Granulariidae Lermontova, 1951
  - Zacanthoididae Swinnerton, 1915
  - Corynexochidae Angelin, 1854
  - Dolichometopidae Walcott, 1916
  - Saukiandidae Hupé, 1953
  - ?Oryctocephalidae Beecher, 1897
    - Oryctocephalinae Beecher, 1897
    - Oryctocarinae Hupé, 1955
    - Lancastriinae Kobayashi, 1935
    - Cheiruroidinae Kobayashi, 1935
    - Tonkinellinae Reed, 1935
- ?Dinesidae Lermontova, 1940

***Order:* Corynexochida Kobayashi, 1935**

***Superfamily:* Corynexochacea Angelin, 1854**

***Family:* Dorypygidae Kobayashi, 1935**

Plate 65. COCAM 1. OLENOIDES Meek, 1877 *Olenoides serratus* Rominger (x2). Burgess shale, Stephen formation, M. Cambrian, British Columbia. (Loaned by MCZ.) Photography of the same pair of trilobites by two approaches: (a) by standard technique; and (b) while immersed in xylene. Note the power of the second method in revealing details of the contour of the carapace. The trilobites from this famous locality are often preserved with soft parts preserved. Unfortunately this specimen does not show the appendages often seen emerging from the edge of the carapace. Compression of the glabella outlines the underlying hypostoma.

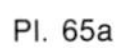

Pl. 65a

Pl. 65b

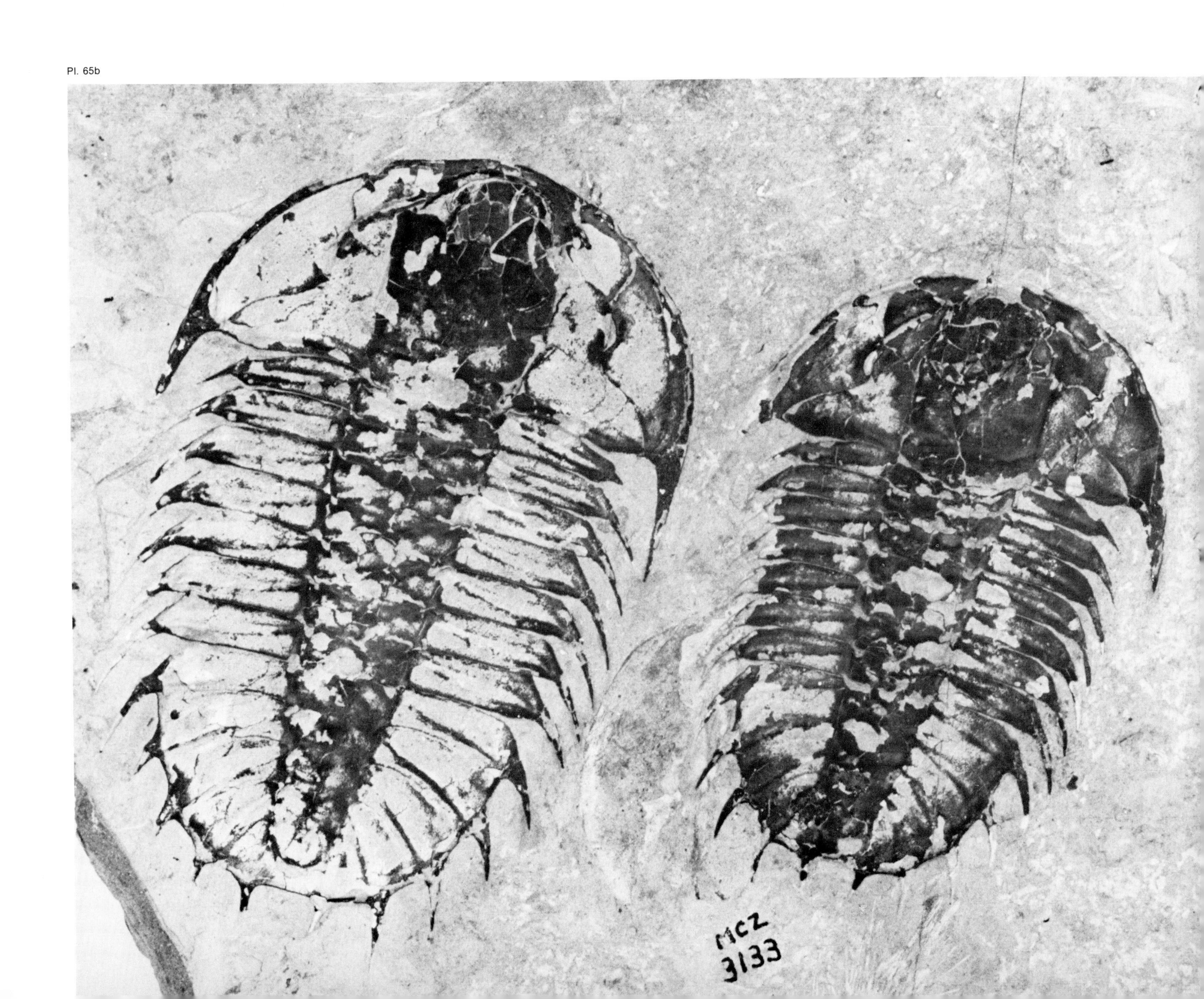

***Family:* Zacanthoididae Swinnerton, 1915**

Plate 66. COCAM 2. ZACANTHOIDES Walcott, 1888. *Zacanthoides typicalis* Walcott (x12). Chisholm shale, M. Cambrian, Half Moon Mine, Pioche, Nevada. Specimen collected by A. Fawcett. (RLS coll., id.) Once again it is worth showing the comparison between standard photography of the specimen in air (a) and what can be seen with xylene immersion (b). The long axial spine originates from the eighth thoracic axial ring.

Pl. 66a

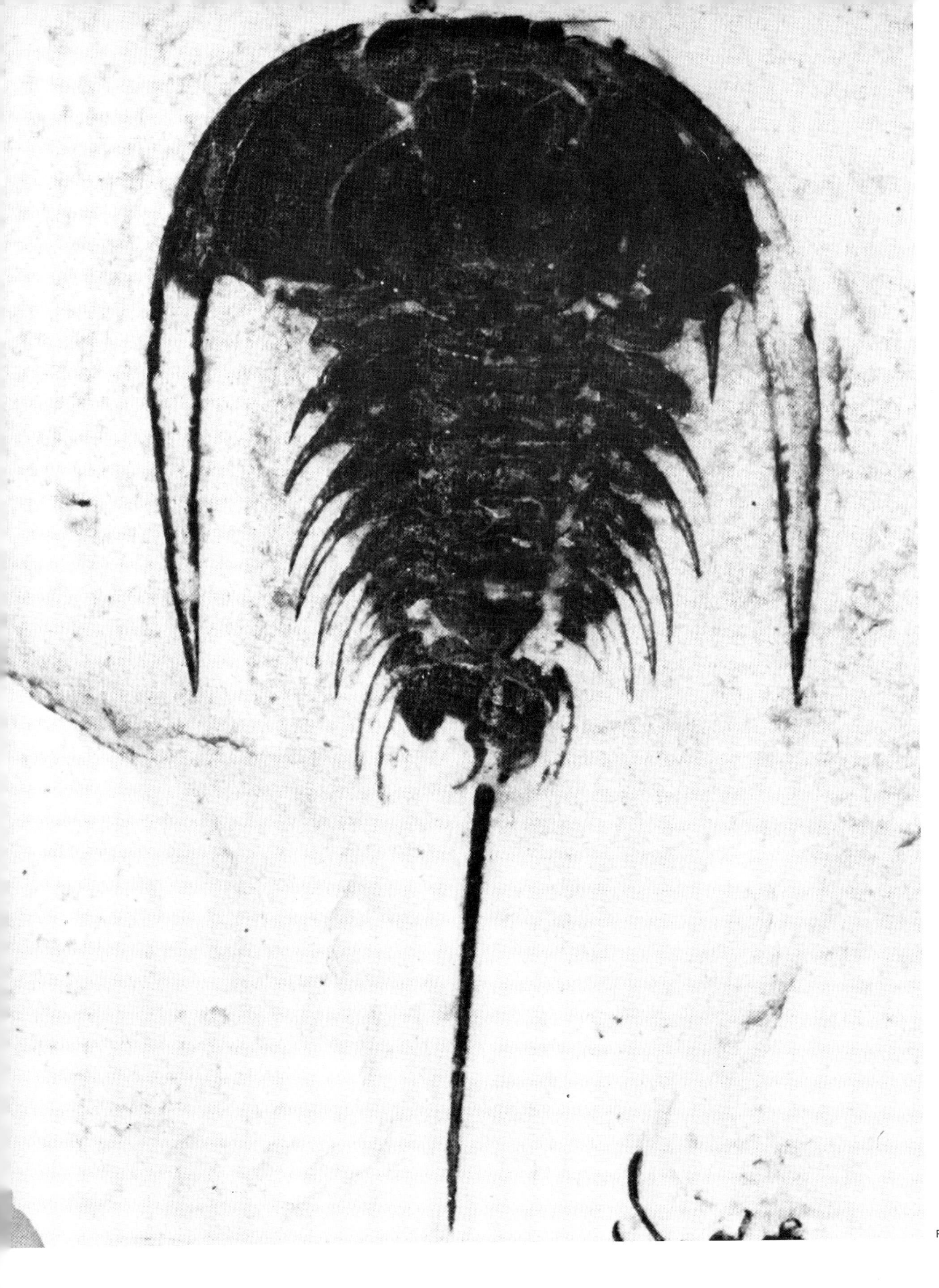

Pl. 66b

Plate 67. COCAM 3. *Zacanthoides typicalis* Walcott (x17.5). Same origin as for COCAM 2. (Specimen collected by A. Fawcett, RLS coll., id.) The trilobite exoskeleton is partly preserved on both sides of the bedding plane. With the aid of xylene immersion, the parts which are missing in (b) can be identified in (a), so that a complete reconstruction can be obtained.

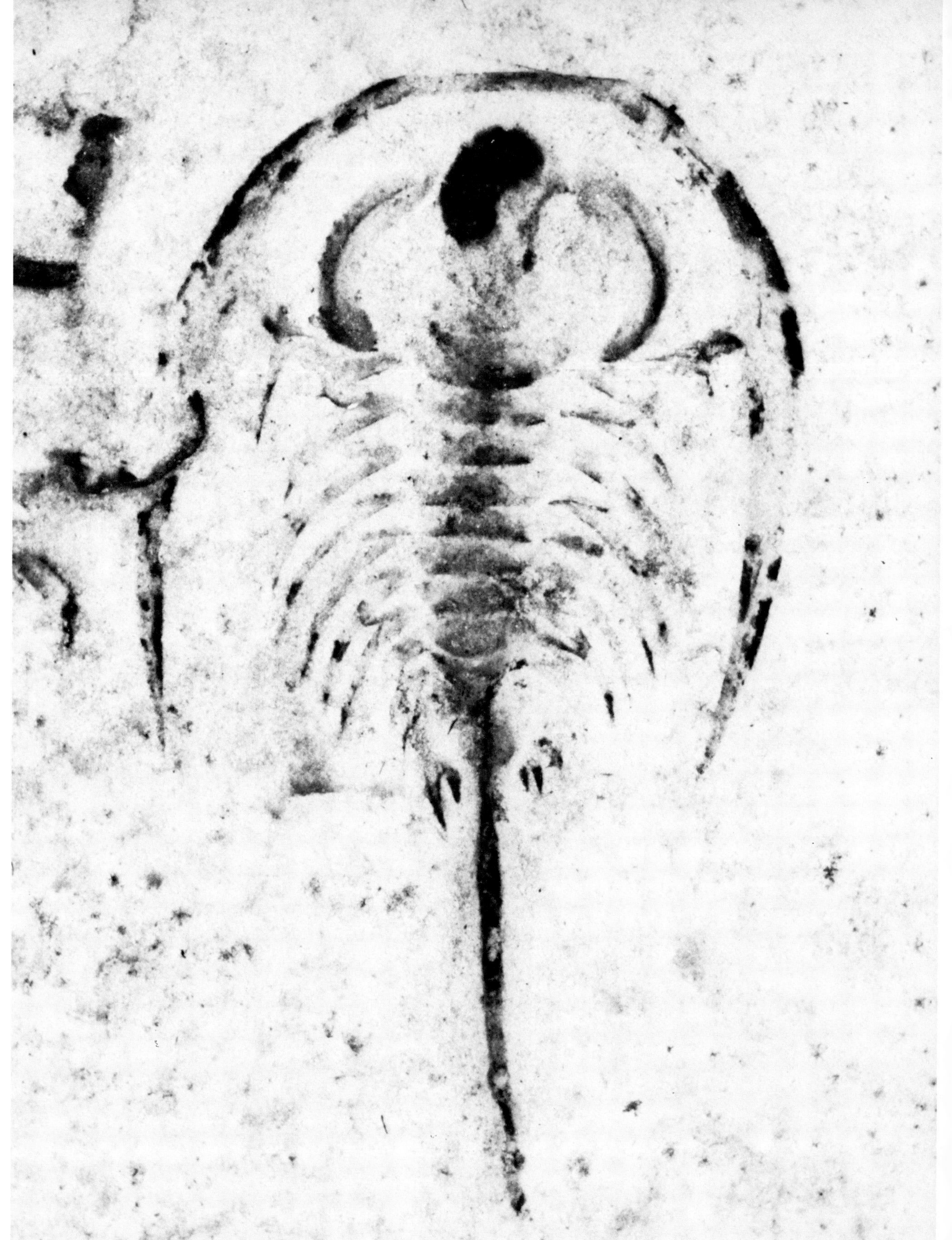

Pl. 67a

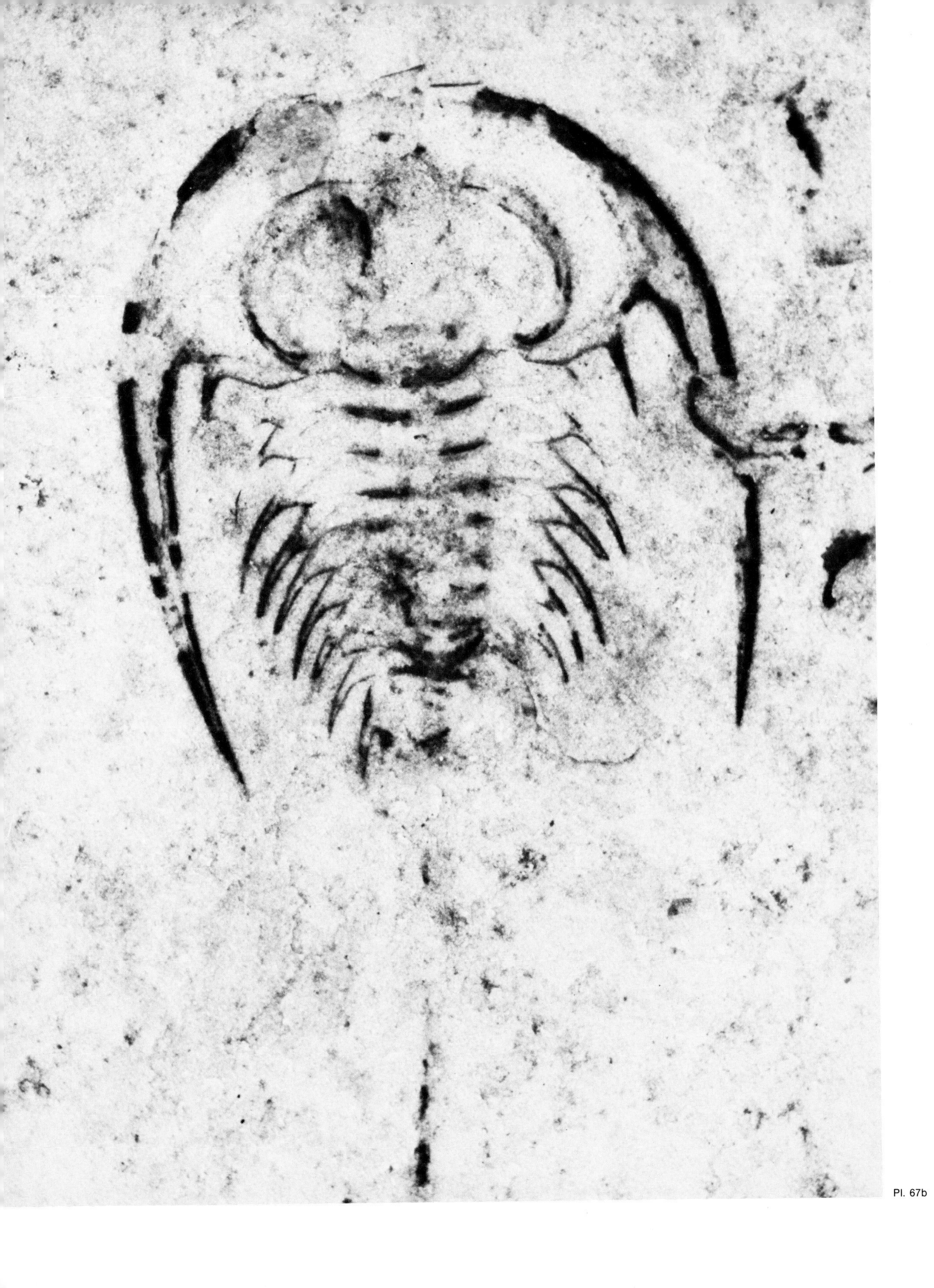

Pl. 67b

Family: **Dolichometopidae Walcott, 1916**

Plate 68. COCAM 4. BATHYURISCUS Meek, 1873. *Bathyuriscus fimbriatus* Robison, (x3.3). Marjum formation, M. Cambrian, between Marjum Pass and Antelope Springs, Millard County, Utah. (Collected by A. Fawcett. RLS coll., id.) Two examples of this trilobite can be seen in this plate, associated with one specimen (in the center) of *Modocia laevinucha* Robison (Family MARJUMIDAE Kobayashi, 1935). Several fragments of agnostid and other trilobites are scattered throughout the plate.

Pl. 68

Plate 69. COCAM 5. Another association of *Bathyuriscus fimbriatus* Robison (top and bottom) with *Modocia laevinucha* Robison (center) (x3). Same origin as for COCAM 4. (Collected by A. Fawcett. RLS coll., Id.)

Pl. 69

Plate 70. COCAM 6. *Bathyuriscus fimbriatus* Robison (x7). Same origin as for COCAM 4. (Collected by A. Fawcett. RLS coll., id.) One of the missing libriginae overlaps, overturned, the right hand side of the thorax, showing prominent radiating alimentary diverticula (prosopon).

Pl. 70

Plate 71. COCAM 7. A complete exoskeleton of *Bathyuriscus fimbriatus* Robison (x13). Same origin as for COCAM 4. (Collected by A. Fawcett. RLS coll., id.)

Pl. 71

Plate 72. COCAM 8. HEMIRHODON Raymond, 1937. *Hemirhodon amplipyge* Robison (x2.3). Marjum fm., Marjum Pass, Millard Co., Utah. (Collected by A. Fawcett. RLS coll., id.) A beautiful example of this large trilobite.

Pl. 72

4.4 Order Illaenida Jaanusson 1959 (nom. trans. ex Illaenina Jaanusson, 1959)

Trilobites generally with terrace lines on doublure, hypostome, and cephalic border; glabella of variable outline; glabellar furrows generally weak or absent; facial suture opisthoparian; rostral plate tends to disappear and librigenae may fuse medially; thorax with comparatively few segments, exceptionally only three; horizontal hingeline generally present; ring joint and dorsal furrow joint may occur, fulcral joint developed as flange joint exceptionally present; panderian organ may occur; pygidium large, generally with entire border; enrollment spheroidal.

Taxa included are:

Superfamily Proetacea Salter, 1864 (emended Bergström, 1973).
—Illaenid trilobites commonly with medium-sized or large eyes; glabella parallel-sided or tapering, generally distinctly delimited in front.

- ?Leiostegiidae Bradley, 1925
  - Leiostegiinae Bradley, 1925
  - Pagodiinae Kobayashi, 1935
- ?Kingstoniidae Kobayashi, 1935
- ?Ptychaspididae Raymond, 1924 (? = Missisquoiidae Hupé, 1953)
- ?Catillicephalidae Raymond, 1938
- ?Eurekiidae Hupé, 1955
- ?Illaenuridae Vogdes, 1890
- ?Shumardiidae Lake, 1907
- Lecanopygidae Lochman, 1953
- Bathyuridae Walcott, 1886
- Holotrachelidae Warburg, 1925
- Proetidae Salter, 1864
  - Proetinae Salter, 1864
  - Cornuproetinae Richter & Richter, 1956
  - Cyrtocymbolinae Hupé, 1955
  - Griffithidinae Hupé, 1955
- Brachymetopidae Prantl & Přibyl, 1950

Superfamily Illaenacea Hawle & Corda, 1847.
—Illaenid trilobites generally with small eyes; glabella generally expanding and faintly delimited anteriorly; pygidial rhachis commonly short; doublure wide.

- Thysanopeltidae Hawle & Corda, 1847 (= Scutelluidae Richter & Richter, 1925)
  - Thysanopeltinae Hawle & Corda, 1847
  - Stygininae Vogdes, 1890
  - Goldillaeninae Balashova, 1959
  - Theamataspidinae Hupé, 1955
- Illaenidae Hawle & Corda, 1847
  - Illaeninae Hawle & Corda, 1847
  - Bumastinae Raymond, 1916
  - Panderiinae Bruton, 1968
  - Ectillaeninae Jaanusson, 1959
- ?Cyclopygidae Raymond, 1925

Section 4.4

Order: **Illaenida Jaanusson, 1959**
Superfamily: **Proetacea Salter, 1864 (emended Bergström, 1973a)**
Family: **Proetidae Salter, 1864**
Subfamily: **Cornuproetinae Richter and Richter, 1956**

Plate 73. ILDEV 1. CORNUPROETUS Richter and Richter, 1919. *Cornuproetus sculptus* (Barrande) (x24). Devonian, Czech. (Negative loaned by Dr. E. N. K. Clarkson, Grant Institute of Geology, Edinburgh, Scotland.) Note the fasciculated surface, origin of the trivial name of this trilobite. The well-developed eyes are considered of the holochroal type.

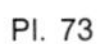

Pl. 73

Subfamily: **Griffithidinae Hupé, 1955**

Plate 74. ILCARB 2. PHILLIPSIA Portlock, 1843. *Phillipsia sampsoni* Vogdes, 1888 (x7.1). Chouteau Limestone, Mississippian, Earley, Boone Co., Missouri. (Loaned by FMNH.) Unusual in this trilobite is the pustulose surface of the carapace. In region of the posterior pleural border in the vicinity of the axial furrows, the pustules degenerates into short spines. Specimen whitened with magnesium oxide.

Pl. 74

Plate 75. ILCARB 3. KASKIA Weller, 1936. *Kaskia chesterensis* Weller (x5.2). U. Mississippian, Crawfordsville, Indiana. (Loaned by FMNH.) Specimen whitened with magnesium oxide.

Pl. 75

***Superfamily:* Illaenacea Hawle and Corda, 1847**

***Family:* Illaenidae Hawle and Corda, 1847**

***Subfamily:* Bumastinae Raymond, 1916**

Plate 76. ILSIL 4. BUMASTUS Murchison, 1839. *Bumastus ioxus* (Hall) (x6.2). Niagaran Limestone, Silurian, Grafton, Illinois. (Gurley coll. of UCWM; loaned by FMNH.) Internal mold, coated with dolomite crystals.

Pl. 76

Plate 77. ILSIL 5. *Bumastus ioxus* (Hall) (xl.5). Rochester shale, Silurian, Lockport, N.Y. (J. Hall coll. of UCWM; loaned by FMNH.) Exoskeleton partially preserved.

4.5. Order Phacopida Salter, 1864 (emended Bergström, 1973a)

Emended diagnosis (Bergström 1973a): Trilobites with long posterior branch of facial suture extending laterally, opisthoparian or proparian; glabella distinctly delimited, slightly narrowing or parallel-sided to expanding forward; generally three pairs of glabellar furrows with deep apodemal invagination some distance from the dorsal furrow; pygidium medium-sized or large, with pleural and interpleural furrows and entire or spinous margin.

Taxa included are:

Superfamily Edelsteinaspidacea Hupé, 1953 (nom. trans., ex Edelsteinaspididae Hupé, 1953).
—Phacopid trilobites with opisthoparian suture; glabella narrowing forwards or parallel-sided, with deep and regular glabellar furrows; ocular ridge present; ventral features poorly known; thorax with more than eleven segments; articulating facets probably not developed.

- Edelsteinaspididae Hupé, 1955
  - Edelsteinaspidinae Hupé, 1955
  - Nodicepinae Suvorova, 1964
  - Laticephalinae Suvorova, 1964

Superfamily Phacopacea Hawle & Corda, 1847 (emended Bergström, 1973a).

—Phacopid trilobites with facial suture proparian or lacking; glabella expanded anteriorly, commonly with somewhat irregularly arranged glabellar furrows; ocular ridge absent; eyes schizochroal; large rostral plate separated from hypostome by suture; thorax with eleven tergites with well-developed articulating facets; enrollment spheroidal.

- Phacopidae Hawle & Corda, 1847
  - Phacopinae Hawle & Corda, 1847
  - Bouleiinae Hupé, 1955
  - Phacopidellinae Delo, 1935
  - Andreaspidinae Struve, 1962
- Pterygometopidae Reed, 1905
  - Pterygometopinae Reed, 1905
  - Chasmopinae Pillet, 1954
- Dalmanitidae Vogdes, 1890
  - Dalmanitinae Vogdes, 1890
  - Zeliszkellinae Delo, 1935
- Calmonidae Delo, 1935
  - Calmoniinae Delo, 1935
  - Acastinae Delo, 1952
- Monorakidae Kramarenko, 1952

Order: **Phacopida Salter, 1864 (emended Bergström, 1973a)**

Superfamily: **Phacopacea Hawle and Corda, 1847 (emended Bergström, 1973a)**

Family: **Phacopidae Hawle and Corda, 1847**

Subfamily: **Phacopinae Hawle and Corda, 1847**

Plate 78. PHSIL 1. EOPHACOPS Delo, 1935. *Eophacops handwerki* Weller (x6.8). Cotype. Niagaran Limestone, Silurian, Lemont, Illinois. (Gift of J. H. Handwerk to UCWM; loaned by FMNH.) Specimen coated with magnesium oxide. The granulose surface is due to incrustation of dolomite crystals.

Pl. 78

Pl. 79a

Pl. 79b

Plate 79. PHDEV 2. PHACOPS Emmrich, 1839. *Phacops rana crassituberculata* Stumm (x3.2). Silica shale, Hamilton group, Devonian (Cazenovian), Sylvania, Ohio. (RLS coll., id.) Side view (a) and top view (b) of a large adult specimen whitened with magnesium oxide. Of all North American trilobites, the *Phacops* of the Silica shale are probably the most spectacular, because of their unusual preservation.

Plate 80. PHDEV 3. *Phacops rana crassituberculata* Stumm (x4). Silica shale, as PHDEV 2. (RLS coll., id.) A perfect specimen, totally extended. The exoskeleton is replaced by dark green calcite, which contrasts against the soft grey shale matrix.

Pl. 80

Plate 81. PHDEV 4. *Phacops rana crassituberculata* Stumm (x5). Silica shale, as for PHDEV 2. (RLS coll., id.) Partially enrolled specimen. The eye structure, clearly visible in this photograph, characterizes this variety of trilobite (see section 3.3).

Pl. 81

Plate 82. PHDEV 5. Group of *Phacops rana* from the Silica shale (x2.3), as for PHDEV 2. The large specimen in upper left corner is an example of *Phacops rana crassituberculata* Stumm, the other three (a small one protrudes in the upper left corner) belong to the variety *Phacops rana milleri* Stewart. The eye structure, as discussed in Section 3.3, distinguishes the latter variety from the former.

Pl. 82

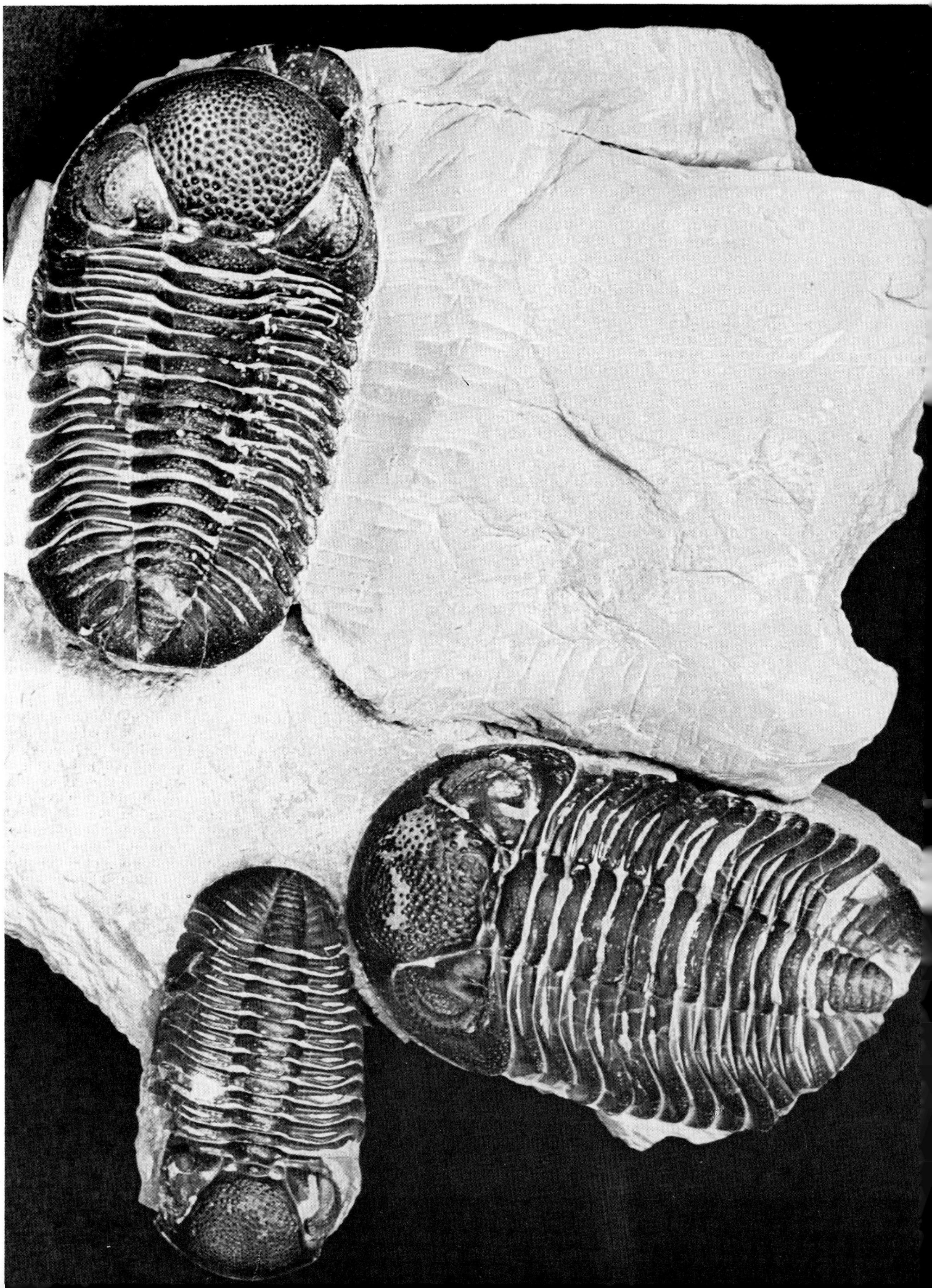

Plate 83. PHDEV 6. A rare group of *Phacops rana milleri* Stewart (x2). Silica shale, as for PHDEV 2. (RLS coll., id.)

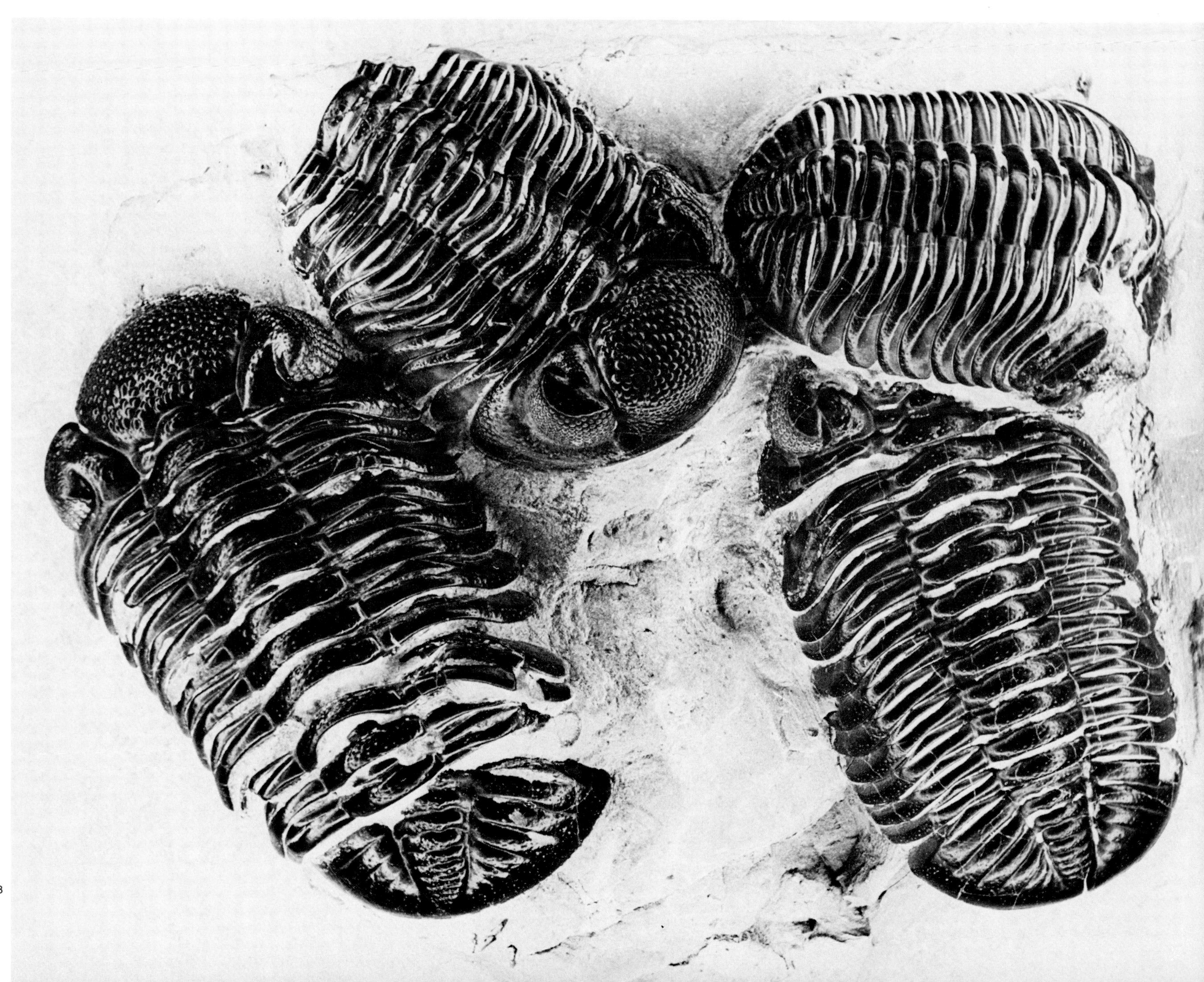

Plate 84. PHDEV 7. *Phacops rana milleri* Stewart (x4.8). Silica shale, as for PHDEV 2. (RLS coll., id.) The same pair of trilobites photographed by two different techniques. In (a) the print is obtained from a color slide; in (b) by standard technique.

Pl. 84a

Pl. 84b

Pl. 85

Plate 85. PHDEV 8. *Phacops rana rana* Green (x2.5). Moscow shale, Devonian, Springbook, Erie Co., N.Y. (Loaned by FMNH.) Print obtained from color slide. Here it is seen that trilobites were gregarious.

Pl. 86

Plate 86. PHDEV 9. Another group of *Phacops rana rana* Green (x3.5). Same origin as for PHDEV 8. (Loaned by FMNH. Print from color slide.)

Plate 87. PHDEV 10. *Phacops latifrons* Bronn (x2.5). Devonian, Rhenish Prussia. (Loaned by FMNH.) Pyritized specimen. Print from color slide.

Pl. 87

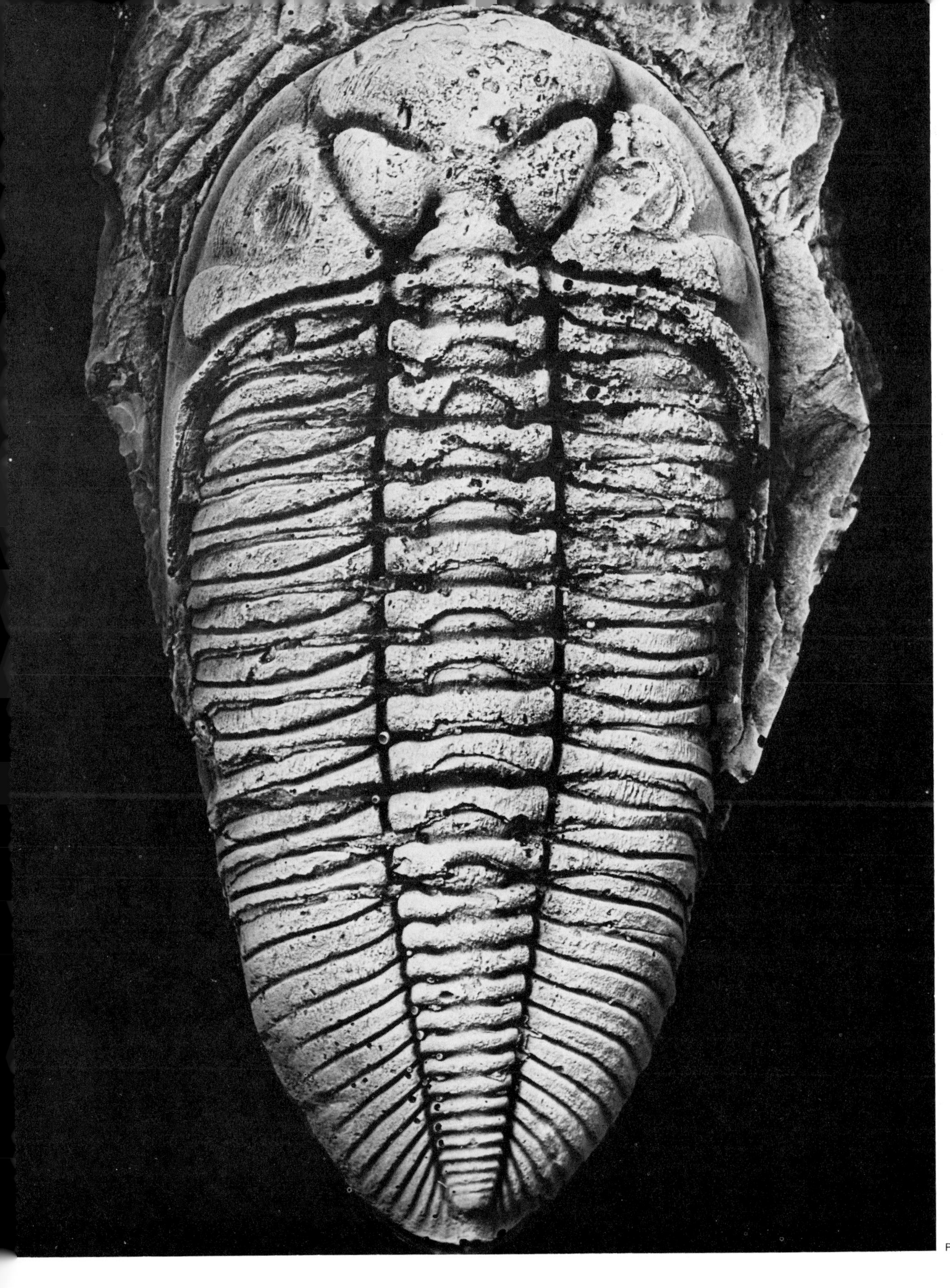

***Family:* Pterygometopidae Reed, 1905**

***Subfamily:* Chasmopinae Pillet, 1954**

Plate 88. PHORD 11. CHASMOPS M'Coy, 1849, *Chasmops extensa* (x2). Ordovician, Oslo region, Norway. (Original at Paleontologisk Museum, University of Oslo.) Cast, whitened with magnesium oxide. The prominent schizochroal eyes of this trilobite (see plate 28 for a congeneric species) are missing in this particular specimen.

Pl. 88

Plate 89. PHSIL 12. DALMANITES Barrande, 1852. *Dalmanites limuloides* (Green) (x3). Rochester shale, Silurian, Lockport, N.Y. (J. Hall coll. of UCWM, loaned by FMNH.) The prominent schizochroal eyes are missing.

Pl. 89

Pl. 90

Plate 90. PHSIL 13. Another specimen of *Dalmanites limuloides* (Green) (x2). Same origin and disposition as PHSIL 12. The eyes seem to be missing in all the specimens from the Rochester shale.

Plate 91. PHSIL 14. *Dalmanites limuloides* (Green) (x1.5). Same origin and disposition as PHSIL 12.

Pl. 91

Plate 92. PHSIL 15. *Dalmanites verrucosus* Hall (x5.1). Waldron shale, Silurian, Waldron, Indiana. (Loaned by FMNH.) Thorax and pygidium disarticulated, seen from the ventral side. Probably exuviae following molting.

Pl. 92

Plate 93. PHSIL 16. *Dalmanites verrucosus* Hall (x6.7). Waldron shale as PHSIL 15. (RLS coll., id.) A small pygidium, showing granules regularly arranged on the axial rings. Disarticulated thoracic segments are scattered throughout the matrix.

Pl. 93

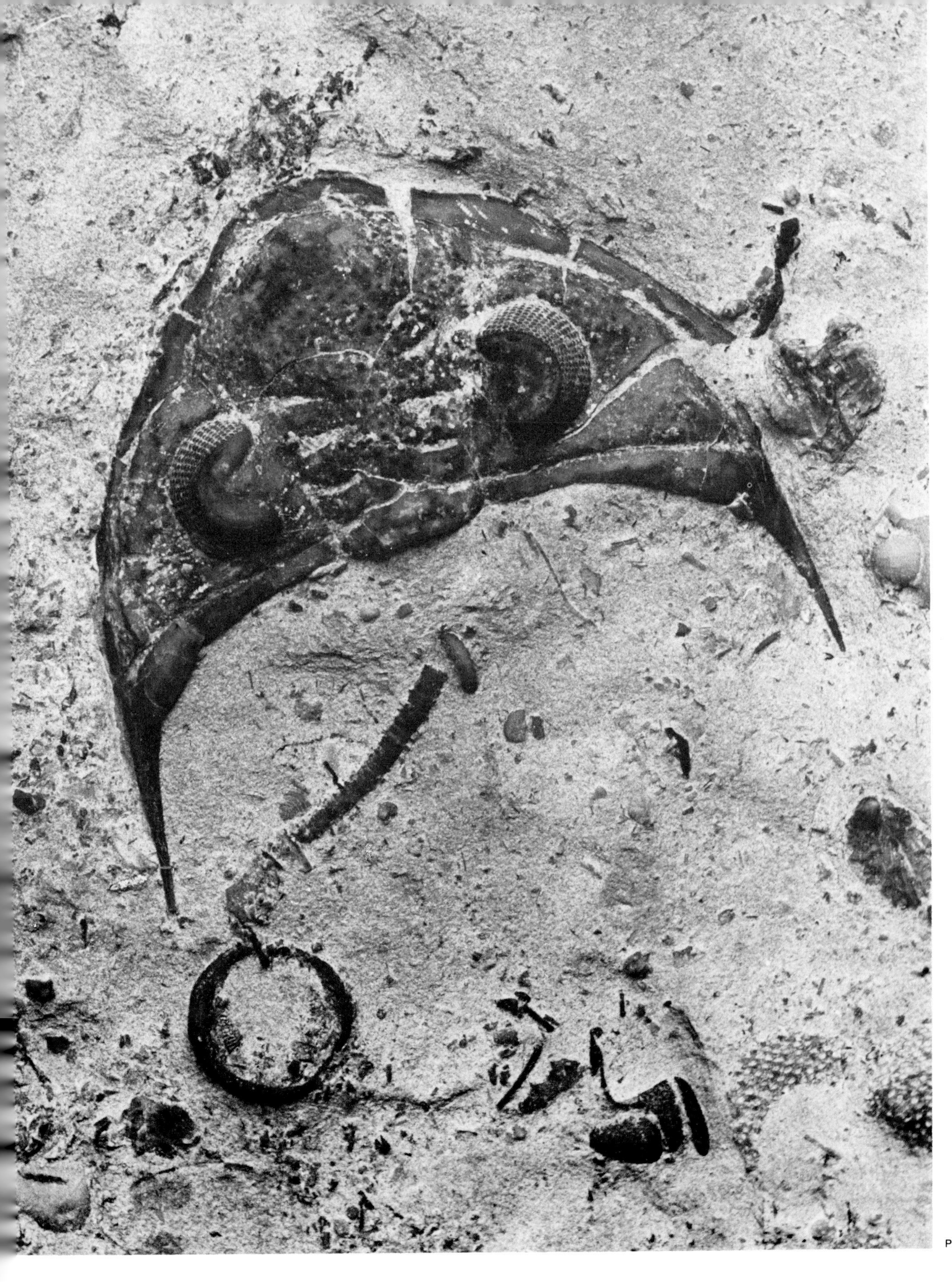

Plate 94. PHSIL 17. *Dalmanites verrucosus* Hall (x4.7). Waldron shale as PHSIL 15. (RLS coll., id.) Cephalon showing the turret-like schizochroal eyes of this trilobite. The shale is impregnated with fossil remains of trilobites and other invertebrates.

Pl. 94

Plate 95. PHSIL 18. *Dalmanites verrucosus* Hall (x3.8). Waldron shale, as for PHSIL 15. (Washburn coll. of UCWM; loaned by FMNH.) Frontal (a) and side view (b) of cephalon, whitened with magnesium oxide.

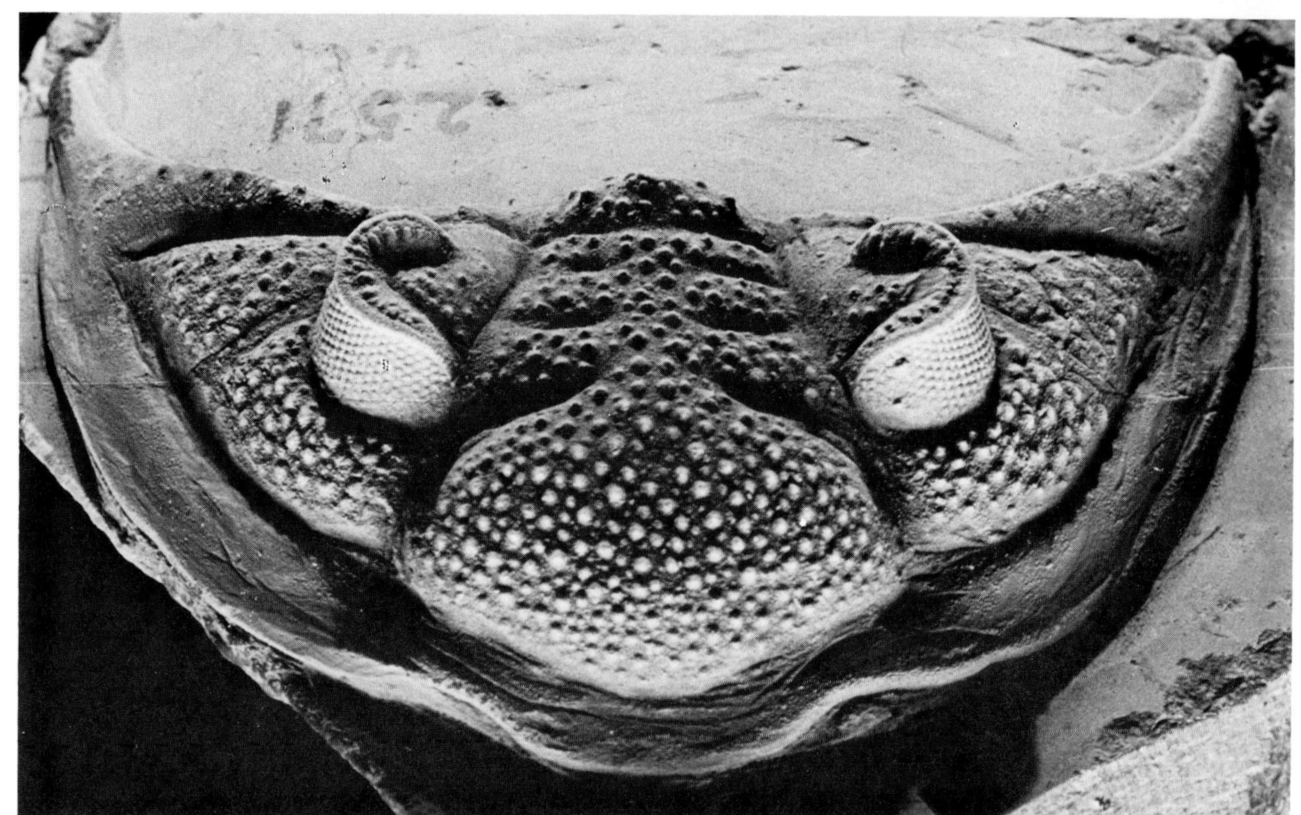

Pl. 95a

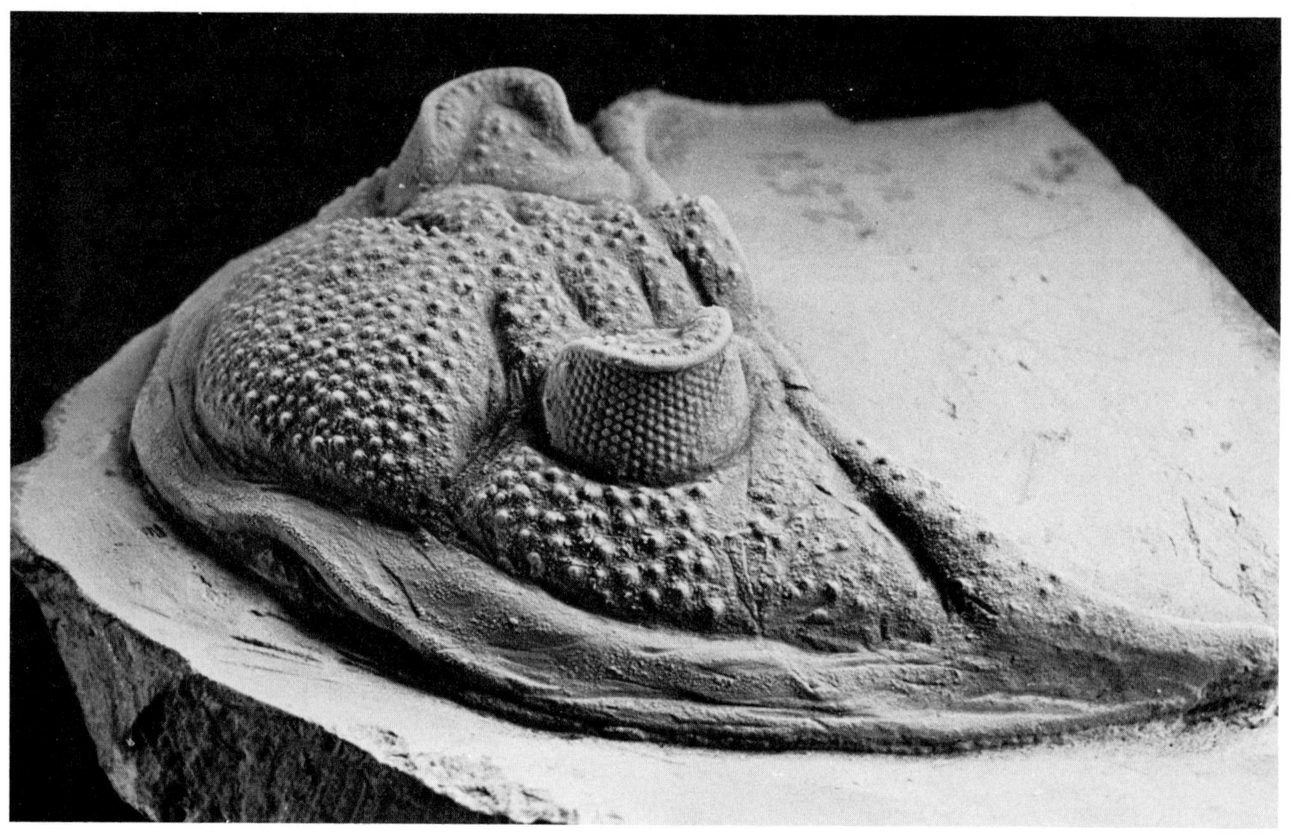

Pl. 95b

Plate 96. PHSIL 19. *Dalmanites platycaudatus* Weller (x3.9). Cotype. Niagaran Limestone, Silurian, Lemont, Illinois. (Collected by L. H. Hyde, UCWM; loaned by FMNH.) Pygidium.

Pl. 96

Plate 97. PHSIL 20. *Dalmanites illinoiensis* Weller (x2.6). Cotype. Niagaran limestone, Silurian, Bonfield, Illinois. (Gurley coll. of UCWM; loaned by FMNH.) Pygidium.

Pl. 97

Subfamily. Zeliszkellinae Struve, 1958

Plate 98. PHORD 21. DALMANITINA Reed, 1905. *Dalmanitina socialis* (Barrande) (xl.5). Letná beds, Ordovician (Caradocian), Veselá, Bohemia. (Loaned by FMNH.) Assemblage of trilobite parts, remarkably well preserved in quartzite. In addition to *Dalmanitina socialis*, two cephala of *Cryptolithus tesselatus* are clearly visible. Due to the high resolution of this photograph, surface details of the carapace as well as the structure of several schizochroal eyes can be identified with the aid of magnifier.

Pl. 98

Plate 99. PHORD 22. *Dalmanitina socialis* (Barrande) (x2). Ordovician, Vraž, Bohemia. (Loaned by FMNH.) The coarser grain of this quartzite, compared with that of PHORD 21, prevents surface details to be discernible.

Pl. 99

Subfamily: **Asteropyginae Delo, 1935**

Plate 100. PHDEV 23. METACANTHINA Pillet, 1954. *Metacanthina barrandei* (Oehlert) (x3.5). Devonian, Oued Ora, Agadir Prov. Morocco. (Gift of Mr. M. Gazay to the author; RLS coll., id.). Specimen weathered from exposure to the extreme climate of the Sahara desert. Identification based on several specimens, including complete cephala, from the same formation.

Pl. 100

Plate 101. PHDEV 24. GREENOPS Delo, 1935. *Greenops boothi* (Green) Windom formation, Devonian, Windom, N.Y. (Loaned by Orton Museum, Ohio State University, courtesy of Stig N. Bergström.) In (a) (x4) the specimen is coated with magnesium oxide. Note how surface details are emphasized. Part (b) (x2.7) shows the uncoated specimen. There is dramatic color contrast, but most of the surface structure is not visible. The eyes in this specimen have been lost.

Pl. 101a

Pl. 101b

Order: Odontopleurida Whittington, 1959 (emended Bergström, 1973a)

Superfamily: **Odontopleuracea Burmeister, 1843**

Family: **Odontopleuridae Burmeister, 1843**

Subfamily: **Odontopleurinae Burmeister, 1843**

4.6. Order Odontopleurida Whittington, 1959 (emended Bergström, 1973a)

Emended diagnosis (Bergström, 1973a): Trilobites generally with strongly marked glabellar furrows, small eyes, surface ornamentation with tubercles or pits but without terrace lines, and opisthoparian to proparian facial suture; postero-laterally directed palpebro-ocular ridge commonly developed; hypostome in contact with rostrum; thorax with entirely encased pleural spines which abut against one another in full enrollment; pleurae generally with imbricating flanges, which extend to or beyond the fulcrum and end with a marginal connective device, either functioning as a fulcral pivot joint or as a limiting device in the enrollment; horizontally directed accessory pleural spines may be present; pygidium commonly fairly small, with marginal spines; enrollment spheroidal or reversed spiral; marginal and pleural spines may be accommodated by vincular groove or pits.

Taxa included are:

Superfamily Odontopleuracea Burmeister, 1843.
—Odontopleurid trilobites with opisthoparian facial sutures; palpebro-ocular ridges well developed; eyes may be on elevated sockets; pleural spines projecting downward; strong accessory pleural spines projecting laterally; pygidium short, with two or three segments.

- Odontopleuridae Burmeister, 1843
  - Odontopleurinae Burmeister, 1843
  - Miraspidinae Richter & Richter, 1947
  - Selenopeltinae Hawle & Corda, 1847
  - Apianurinae Whittington, 1956
- Eoacidaspididae Poletaeva, 1957
- ?Glaphuridae Hupé, 1955

Superfamily Cheiruracea Salter (emended Bergström, 1973).
—Odontopleurid trilobites generally with proparian or gonatoparian facial suture; no accessory pleural spines; pygidium short or long.

- ?Telephinidae Marek, 1952
- ?Raymondinidae Clark, 1924
- Carmonidae Kielan, 1960
- Celmidae Jaanusson, 1956
- Hammatocnemidae Kielan, 1960
- Cheiruridae Salter, 1864
  - Cheirurinae Salter, 1864
  - Eccoptochilinae Lane, 1971
  - Sphaerexochinae Öpik, 1937
  - Deiphoninae Raymond, 1913
  - Acanthoparyphinae Whittington & Evitt, 1953
  - Pilekiinae Sdzuy, 1955
- Pliomeridae Raymond, 1913
  - Pliomerinae Raymond, 1913
  - Placopariinae Hupé, 1955
  - Pliomerellinae Hupé, 1955
  - Diaphanometopinae Jaanusson, 1959
- Encrinuridae Angelin, 1854
  - Encrinurinae Angelin, 1854
  - Cybelinae Holiday, 1942
  - Dindymeninae Henningsmoen, 1959
  - Staurocephalinae Prantl & Přibyl, 1947

Plate 102. ODSIL 1. ODONTOPLEURA Emmrich, 1839. *Odontopleura ovata* Emmrich (x10). Motol beds, M. Silurian (Wenlock), Loděnice, Bohemia. (RLS coll., id.; courtesy of MCZ.) The trilobites in this sequence from Bohemia are very striking for their delicate and extravagant morphology. The long arching spines are indeed pleural or genal or pygidial spines, an integral part of the exoskeleton, and do not represent appendages or walking legs. Two cephala of *Aulacopleura koninckii koninckii* (Barrande) are visible in the upper right side of the photograph.

Pl. 102

Plate 103. ODSIL 2. *Odontopleura ovata* Emmrich (x5). Same origin as for ODSIL 1. (Loaned by MCZ; RLS id.) In spite of the convex appearance, this is actually the external (concave) impression of the trilobite in the shale. Part of the exoskeleton remained attached to the mold.

Pl. 103

Subfamily: Miraspidinae Richter and Richter, 1917

Plate 104. ODSIL 3. MIRASPIS Richter and Richter, 1917. *Miraspis mira* (Barrande) (x9). Motol beds, M. Silurian (Wenlock), Loděnice, Bohemia. (RLS coll., id.; courtesy MCZ). Each pleural segment carries two kinds of spines: one is long and slender; the other, shorter, is laced with secondary spines. This is possibly the most photogenic trilobite of this entire collection.

Pl. 104

Plate 106. ODSIL 5. *Miraspis mira* (Barrande) (x5.7). Same origin as for ODSIL 3. (RLS coll., id.; courtesy MCZ.)

Plate 105. ODSIL 4. Another example of *Miraspis mira* (Barrande) (x7.5) from the same locality as ODSIL 3. (Loaned by MCZ; RLS id.)

Pl. 105

Pl. 106

Plate 107. ODSIL 6. *Miraspis mira* (Barrande) (x10). Same origin as for ODSIL 3. (Loaned by MCZ.) This is the only specimen figured here in which the peduncles, or stalks, carrying the eyes at their tips are preserved.

Pl. 107

Superfamily: **Cheiruracea Salter (emended Bergström, 1973)**

Family: **Cheiruridae Salter, 1864**

Plate 108. ODSIL 7. CHEIRURUS Beyrich, 1845. *Cheirurus dilatatus* Raymond (x6.1) Waldron shale, Silurian, Waldron, Indiana. (RLS coll., id.) Cranidium of this rare trilobite.

Pl. 108

Plate 109. ODSIL 8. *Cheirurus hydei* (Weller) (x6.3) Holotype. Niagaran Limestone, Silurian, Lemont, Illinois. (Collected by L. H. Hyde. UCWM coll.; loaned by FMNH.)

Pl. 109

Plate 110. ODSIL 9. *Cheirurus hydei* (Weller) (negative mold confronting two specimens of *Calymene celebra* Raymond (x3). Niagaran formation, Silurian, Grafton, Illinois. (RLS coll., id.) Although an external impression, this example of *Cheirurus hydei* exhibits better details than the holotype, in particular the genal and pygidial spines and the proparian suture.

Pl. 110

Pl. 111

***Subfamily:* Sphaerexochinae Öpik, 1937**

Plate 111. ODSIL 10. SPHAEREXOCHUS Beyrich, 1845. *Sphaerexochus romingeri* Hall (x2.3). Niagaran group, Silurian, Racine, Wisconsin. (Van Horne coll. of UCWM. Loaned by FMNH.) Several cranidia, in different orientation, of this peculiar trilobite. Only the lowermost is a complete cephalon, showing cephalic suture and eye lobe.

Pl. 112

Plate 112. ODSIL 11. *Sphaerexochus romingeri* Hall (x4). Same origin as for ODSIL 10. Enlarged top view of cranidium.

Family: **Encrinuridae Angelin, 1854**

Plate 113. ODSIL 12. ENCRINURUS Emmrich, 1844. *Encrinurus egani* Miller (x6.4). Plesiotype. Niagaran group, Silurian, Lemont, Illinois. (Van Horne coll. of UCWM; loaned by FMNH.) Specimen whitened with magnesium oxide. The eyes, which would appear on short peduncles, are missing.

Pl. 113

Plate 114. ODSIL 13. *Encrinurus egani* Miller (x5.8). Niagaran group, Silurian, Racine, Wisconsin. (UCWM coll.; loaned by FMNH.) Complete specimen, transversly compressed.

Pl. 114

Plate 115. ODSIL 14. *Encrinurus egani* Miller (x5.1), as in ODSIL 12 and 13. (James coll. of UCWM; loaned by FMNH.) Side view of complete individual, showing the erect thoracic axial spine. Facial suture clearly visible.

Pl. 115

Plate 116. ODSIL 15. *Encrinurus indianensis* Kindle and Breger (x6.1). Niagaran group, Silurian, Burlington, Wisconsin. (UCWM coll.; loaned by FMNH.) External impression (negative mold) encrusted with dolomite crystals. The cephalon is considerably more tuberculate than in *Encrinurus egani*.

Pl. 116

4.7. Order Lichida Moore, 1959

Medium-sized to large trilobites, generally with tuberculate ornament; glabella with longitudinally elongated glabellar furrows; eyes small; facial suture opisthoparian; hypostome subquadrate, fixed to unpaired rostrum, thorax with horizontal hingeline short or missing; flanges, ring joint, dorsal furrow joint, fulcral joint, and panderian organ not developed; pygidium large with generally three pairs of pleural segments; probably not enrolling.

Taxa included are:

Superfamily Lichacea Hawle & Corda, 1847
- Lichidae Hawle & Corda, 1847
  - Lichinae Hawle & Corda, 1847
  - Homolichinae Phleger, 1936
  - Tetralichinae Phleger, 1936
  - Ceratarginae Tripp, 1957
- Lichakephalidae Tripp, 1957

Order: **Lichida Moore, 1959**

Superfamily: **Lichacea Hawle and Corda, 1847**

Family: **Lichidae Hawle and Corda, 1847**

Subfamily: **Lichinae Hawle and Corda, 1847**

Plate 117. LISIL 1. ARCTINURUS Castelnau, 1843. *Arctinurus occidentalis* Hall, (x3.3). Waldron shale, Silurian, Waldron, Indiana. (Washburn coll. of UCWM; loaned by FMNH.) A complete pygidium.

Pl. 118

Plate 118. LISIL 2. *Arctinurus occidentalis* Hall (x1.45). Rochester shale, Silurian, Lockport, N.Y. (UCWM coll.; loaned by FMNH.) Specimen whitened with magnesium oxide. Part of the glabella missing.

4.8. Order Ptychopariida Swinnerton, 1915 (emended Bergström, 1973a)

Emended diagnosis (Bergström, 1973a): small to medium-sized, seldom large trilobites, as a rule with opisthoparian or gonatoparian suture; compound eyes generally small or absent; glabella commonly trapezoidal, with glabellar furrows undifferentiated or absent; hypostome either not in contact with rostrum or with a transverse depression for the reception of the pygidial margin in the enrolled trilobite; thorax with blunt pleural spines; horizontal hinge line present in most members, rarely secondarily lost; flanges only exceptionally developed; ring joint commonly present, dorsal furrow joint and fulcral joint usually absent; pygidium generally small to medium-sized, with entire border, which may be geniculated; spiral enrollment or derived type.

Taxa included are:

Superfamily Solenopleuracea Angelin, 1854 (emended Bergström, 1973a)

—Ptychopariid trilobites generally with tapering glabella; palpebro-ocular ridges commonly visible; commonly long thorax and short pygidium; generally typical or unrolled spiral enrollment.

- Ellipsocephalidae Matthew, 1887
  - Ellipsocephalinae Matthew, 1887
  - Strenuellinae Hupé, 1953
  - Antatlasiinae Hupé, 1953
  - Kingaspidinae Hupé, 1953
  - Agraulinae Raymond, 1913
- Ptychopariidae Matthew, 1887
  - Ptychoparinae Matthew, 1887
  - Antagminae Hupé, 1953
  - ?Nassoviinae Howell, 1937
- Solenopleuridae Angelin, 1854
  - Solenopleurinae Angelin, 1854
  - Acrocephalitinae Hupé, 1953
  - Saoinae Hupé, 1953
  - Hystricurinae Hupé, 1953
  - Dimeropyginae Hupé, 1953
- Conocoryphidae Angelin, 1854
  - Conocoryphinae Angelin, 1854
  - Pharostomatinae Hupé, 1953
  - Periommellinae Rasetti, 1955
- Bolaspididae Howell, 1959
- Nepeidae Whitehouse, 1939
- Menomoniidae Walcott, 1916
- Aulacodigmatidae Öpik, 1967
- Eulomatidae Kobayashi, 1955
- Plethopeltidae Raymond, 1924
- Harpidae Hawle & Corda, 1847
- Entomaspididae Ulrich in Bridge, 1930
- Aulacopleuridae Angelin, 1854
  - Aulacopleurinae Angelin, 1854
  - Otarioninae Richter & Richter, 1926
  - Cyphaspidinae Přibyl, 1947
- Phillipsinellidae Whittington, 1950

Superfamily Calymenacea Burmeister, 1843

—Ptychopariid trilobites with trapezoidal glabella; palpebro-ocular ridges usually absent; facial sutures generally gonatoparian; hypostome fixed to rostrum; pygidium generally long and triangular.

- Calymenidae Burmeister, 1843
- Homalonotidae E. J. Chapman, 1890
  - Homalonotinae E. J. Chapman, 1890
  - Portagininae Lespérance, 1968
  - Bavarillinae Sdzuy, 1957
  - Eohomalonotinae Hupé, 1955
  - Colpocoryphinae Hupé, 1955

Superfamily Trinucleacea Hawle & Corda, 1847

—Ptychopariid trilobites generally with forward-expanding and convex glabella; paired eyes generally absent; facial suture opisthoparian, submarginal or marginal anteriorly and anterolaterally; librigena in one piece, rostrum absent; genal spine long; thorax generally short and flat; generally basket and lid enrollment.

- Trinucleidae Hawle & Corda, 1847
  - Trinucleinae Hawle & Corda, 1847
  - Tretaspidinae Whittington, 1941
  - Cryptolithinae Angelin, 1854
  - Novaspidinae Whittington, 1941
  - Incaiinae Huges & Wright, 1970
- Orometopidae Hupé, 1955
- Dionididae Gürich, 1908
- Raphiophoridae Angelin, 1854
  - Raphiophorinae Angelin, 1854
  - Ampyxininae Hupé, 1955
  - Ampyxellinae Koroleva, 1959
  - Bulbaspidinae Kobayashi & Hamada, 1971
- Endymioniidae Raymond, 1920
- Alsataspididae Turner, 1940
- Hapalopleuridae Harrington & Leanza, 1957
- Myindidae Hupé, 1955

Order: **Ptychopariida Swinnerton, 1915 (emended Bergström, 1973a)**

Superfamily: **Solenopleuracea Angelin, 1854 (emended Bergström, 1973a)**

Family: **Ellipsocephalidae Matthew, 1887**

Subfamily: **Ellipsocephalinae Matthew, 1887**

Plate 119. PTCAM 1. ELLIPSOCEPHALUS Zenker, 1833. *Ellipsocephalus hoffi* (Schlotheim) (x2). M. Cambrian, Jinetz, Bohemia. (Loaned by Geological Enterprises, Ardmore, Oklahoma.) This trilobite was blind.

Pl. 119

Pl. 120

Plate 120. PTCAM 2. CONO-CORYPHE Hawle and Corda, 1847. *Conocoryphe sulzeri* Schlotheim (x2). M. Cambrian (Jinetz shale), Beroun near Prag, Bohemia. (Loaned by MCZ.) Another blind trilobite. The specimen on the lower left-hand corner is a negative mold, the other two are positive casts (steinkerns).

Plate 121. PTCAM 3. *Conocoryphe sulzeri* Schloteim as for PTCAM 2 (x3.8). (RLS coll.; courtesy of MCZ.) What black-and-white photography cannot convey is, of course, the color of some of the trilobite specimens. In this case the trilobite is coated by a bright yellow-ochre film of limonite which contrasts with the tan matrix background.

Pl. 121

***Family:* Menomoniidae Walcott, 1916**

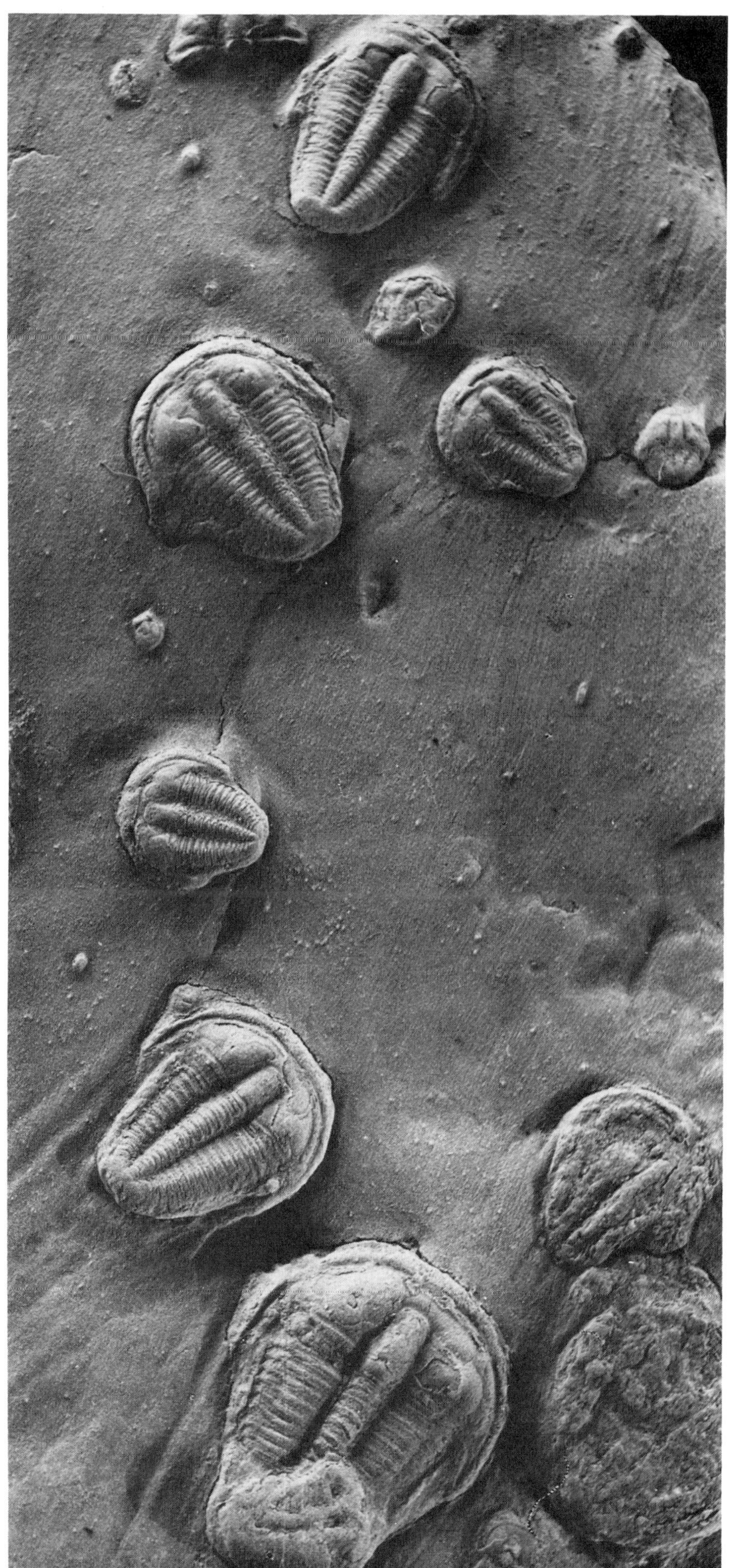

Pl. 122

Plate 122. PTCAM 4. BOLASPIDELLA Resser, 1937. *Bolaspidella housensis* (Walcott) (x5.2). M. Cambrian, Wheeler formation, Antelope Springs, Utah. (Collected by A. Fawcett. RLS coll., id.) Specimen whitened with magnesium oxide. On the uncoated specimen the trilobites appear black on a light grey soft matrix.

***Family:* Aulacopleuridae Angelin, 1854**

***Subfamily:* Aulacopleurinae Angelin, 1854**

Plate 123. PTSIL 5. AULACO-PLEURA Hawle and Corda, 1847. *Aulacopleura koninckii koninckii* (Barrande) (x9). M. Silurian (Wenlock), Beroun near Prag, Bohemia. (RLS coll., courtesy MCZ.)

Pl. 123

Plate 124. PTSIL 6. Another example of *Aulacopeura koninckii koninckii* (Barrande) (x7), as in plate 123. (RLS coll., courtesy MCZ.)

Pl. 124

Superfamily: **Calymenacea Burmeister, 1843**

Family: **Calymenidae Burmeister, 1843**

Plate 125. PTORD 7. FLEXICALYMENE Shirley, 1936. *Flexicalymene meeki* (Foerste) (x5.3). Waynesville Formation, Fort Ancient Member, U. Ordovician, Westwood, Cincinnati, Ohio. (RLS coll.) Specimen whitened with magnesium oxide. The entire exoskeleton is preserved here, replaced by calcite.

Pl. 125

Pl. 126a

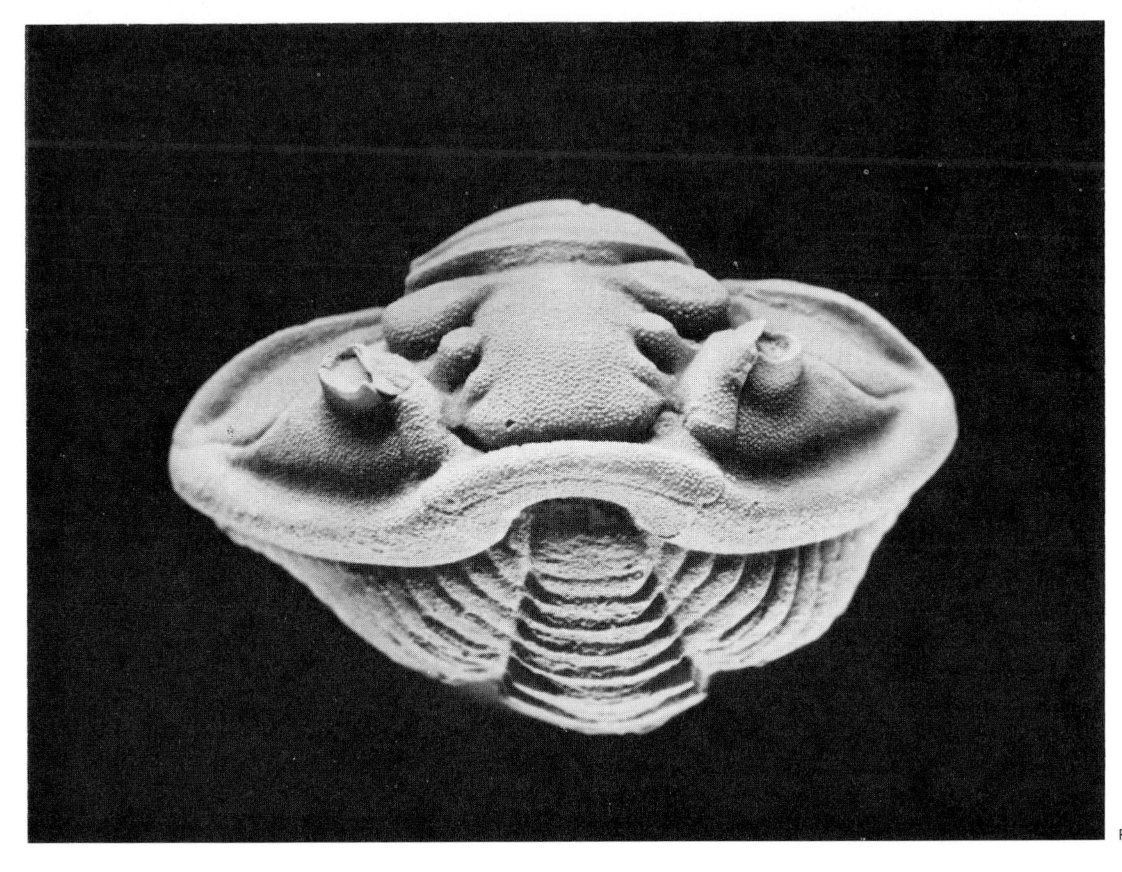

Pl. 126b

Plate 126. PTORD 8. *Flexicalymene meeki* (Foerste) (x5.5) as in plate 125. Enrolled specimen collected at Waynesville, Ohio (RLS coll., id.) An enlarged profile view of the same specimen is shown in plate 35 as example of "uncoiled spiral enrollment." Note papillate or granular surface of carapace, emphasized by magnesium oxide coating.

Plate 127. PTSIL 9. CALYMENE Brongniart, 1822. *Calymene breviceps* Raymond (x9.4). Waldron shale, Silurian, Niagaran, Waldron, Indiana. (RLS coll., id.) The exoskeleton is replaced by calcite. Specimen collected by Susanna Ginali.

Pl. 127

Pl. 128a

Pl. 128b

Pl. 128c

Pl. 128d

Plate 128. PTSIL 10. *Calymene breviceps* Raymond, as in PTSIL 9 (x2.9). (Gurley coll. of UCWM; loaned by FMNH.) Several complete individuals, rolled and unrolled.

Plate 129. PTSIL 11. *Calymene niagarensis* Hall (x3.1). Racine Formation in the Niagaran Limestone, Silurian, McCook, Illinois (RLS coll., id.) Negative mold encrusted with dolomite crystals. In spite of the coarse grain of the surface, anatomical details of the exoskeleton are remarkably well preserved in this kind of fossilization. In this and other similar specimens collected by the author at the above locality, a hollow cavity is found in place of the steinkern.

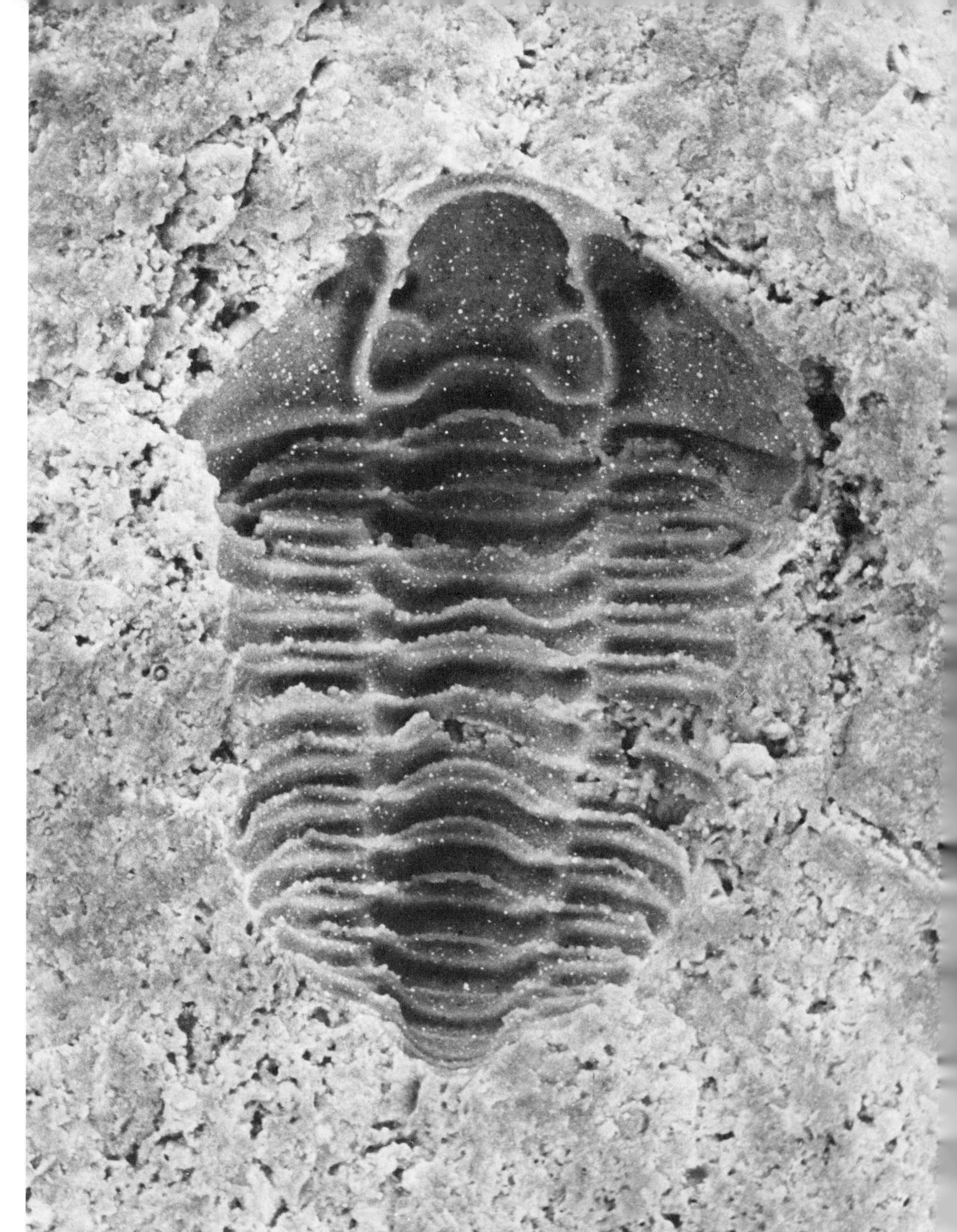

Pl. 129

Pl. 130

Plate 130. PTSIL 12. *Calymene celebra* Raymond (x3.5). Joliet Formation in the Niagaran Limestone, Silurian, Lehigh, Illinois. (RLS coll., id.) Steinkern and its negative mold, specimen collected by Matteo Levi-Setti, age 4.

Plate 131. PTSIL 13. Side view of another example of *Calymene celebra* Raymond (x5.1), from Lehigh, Illinois. (RLS coll., id.) The layer previously occupied by the exoskeleton is replaced by dolomite crystals. Specimen collected by Emile Levi-Setti, age 7.

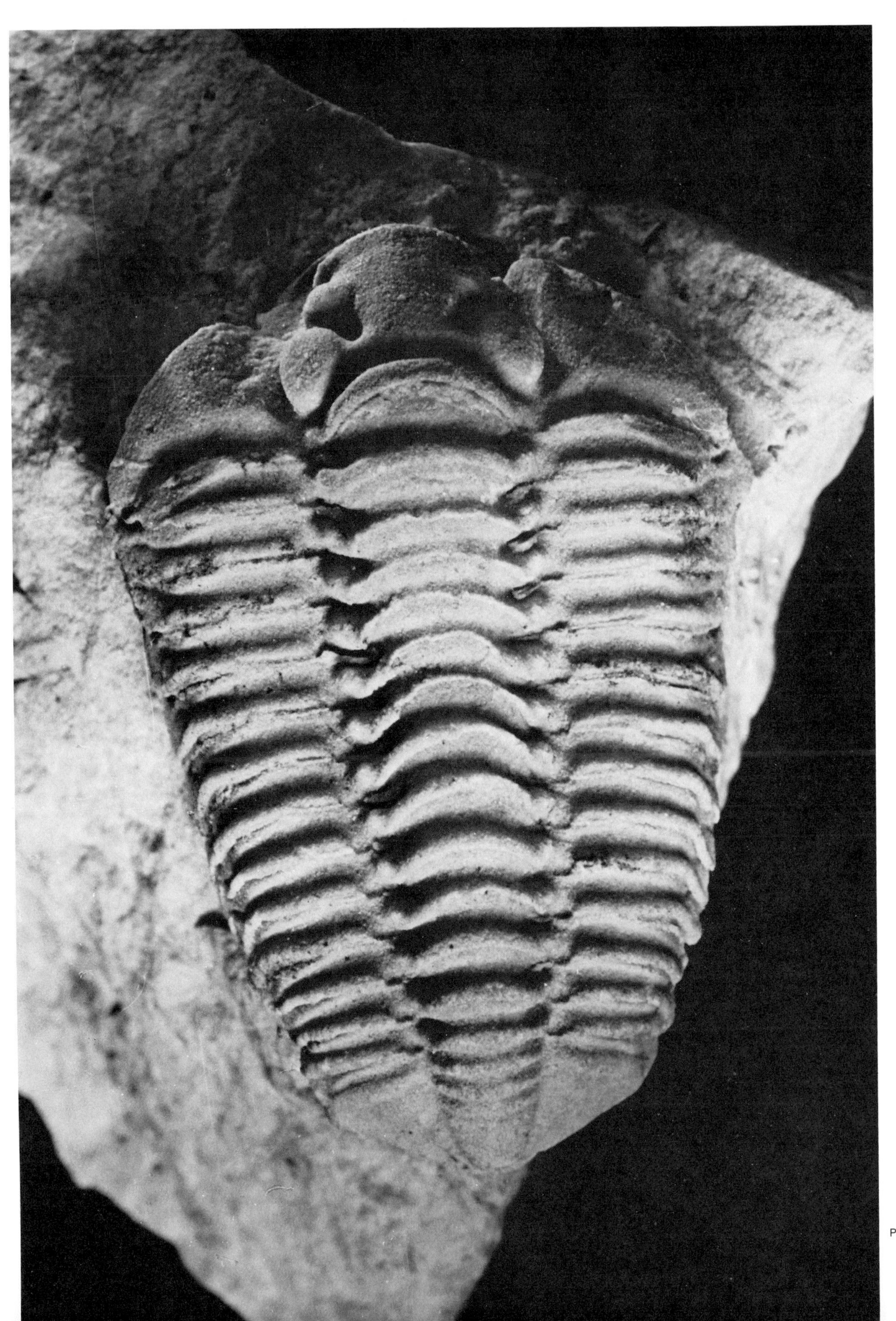

Plate 132. PTSIL 14. *Calymene celebra* Raymond (x6.5). Niagaran Limestone, Silurian, Milwaukee, Wisconsin. (RLS coll., id.)

Pl. 132

Plate 133. PTSIL 15. Negative mold of several *Calymene* individuals (x1.2). Niagaran Limestone, Silurian, Wisconsin. (Loaned by FMNH.)

Pl. 133

Plate 134. PTSIL 16. *Calymene celebra* Raymond (x3.8). Niagaran limestone, Silurian, Grafton, Illinois. (RLS coll., id.)

Pl. 134

Pl. 135

Plate 135. PTSIL 17. *Calymene celebra* Raymond (x4). Same origin as for PTSIL 16. (UCWM coll.; loaned by FMNH.) An unusual association of complete individuals.

Pl. 136

Plate 136. PTSIL 18. DIACALYMENE Kegel, 1927. *Diacalymene clavicula* (Campbell) (x2.2). Henryhouse Formation, Hunton Group, Silurian (Cayugan). Lawrence uplift near Ada, Oklahoma. (RLS coll.)

Plate 137. PTSIL 19. *Diacalymene clavicula* (Campbell) as for PSIL 18 (x2.5). (RLS coll.) The exoskeleton is replaced by calcite; the matrix is a soft siltstone. Several fractures due to compression in the sediment can be noticed.

Pl. 137

Family: **Homalonotidae E. J. Chapman, 1890**

Subfamily: **Homalonotinae E. J. Chapman, 1890**

Plate 138. PTSIL 20. TRIMERUS Green, 1832. *Trimerus delphinocephalus* Green (x7.5). Rochester shale, M. Silurian Lockport, N.Y. (J. Hall coll., UCWM; loaned by FMNH; RLS id.)

Pl. 138

Superfamily: **Trinucleacea Hawle and Corda, 1847**

Family: **Trinucleidae Hawle and Corda, 1847**

Subfamily: **Cryptolithinae Angelin, 1854**

Plate 139. PTORD 21. CRYPTOLITHUS Green, 1832. *Cryptolithus tesselatus* Green (x2.8). Pulaski shale, M. Ordovician, Lorain, Jefferson Co., N.Y. (J. Hall coll., UCWM; loaned by FMNH.) Specimen whitened with magnesium oxide.

Family: **Radiophoridae Angelin, 1854**

Subfamily: **Ampyxininae Hupé, 1955**

Plate 140. PTORD 22. AMPYXINA Ulrich, 1922. *Ampyxina bellatula* (Savage) (x2). Maquoketa Formation, U. Ordovician, near Elsberry, Missouri. (RLS coll., id.)

Pl. 140

4.9. Order Agnostida Salter, 1864 (=Miomera Jaekel, 1909)

Small trilobites with thorax consisting of two or three segments; generally without dorsal eyes and then lacking ecdysial facial sutures; horizontal hingeline developed; intestinal diverticula excessively developed, commonly traceable in exterior reticulation, filling genal regions.

Taxa included are:

Superfamily Agnostacea M'Coy, 1849.

—Agnostid trilobites without facial sutures and dorsal eyes; two thoracic segments with rhachis distinctly different from pygidial rhachis; rhachis on pygidium with a maximum of three well defined segments, but segments commonly not distinguishable.

- Agnostidae M'Coy, 1849
  - Agnostinae M'Coy, 1849
  - Ptychagnostinae Kobayashi, 1939
  - Quadragnostinae Howell, 1935
- Diplagnostidae Whitehouse, 1936
  - Diplagnostinae Whitehouse, 1936
  - Oidalagnostinae Öpik, 1967
  - Tomagnostinae Kobayashi, 1940
  - Ammagnostinae Öpik, 1967
  - Pseudagnostinae Whitehouse, 1936
  - Glyptagnostinae Whitehouse, 1936
- Clavagnostidae Howell, 1937
  - Clavagnostinae Howell, 1937
  - Aspidagnostinae Pokrovskaja, 1960
- Trinodidae Howell, 1935 (= Geragnostidae Howell, 1935)
- Discagnostidae Öpik, 1963
- Sphaeragnostidae Kobayashi, 1939
- Phalacromidae Hawle & Corda, 1847
- ?Condylopygidae Raymond, 1913
- Spinagnostidae Howell, 1935

Superfamily Eodiscacea Raymond, 1913 (nom. transl., ex Eodiscidae Raymond, 1913).

—Agnostid trilobites with dorsal eyes and proparian facial sutures, or without dorsal eyes and facial sutures; thoracic and pygidial rhachis of similar morphology; long pygidial rhachis generally with more than three distinctly visible segments.

- Eodiscidae Raymond, 1913
- Pagetiidae Kobayashi, 1935

Order: **Agnostida Kobayashi, 1935**
Family: **Agnostidae McCoy, 1849**

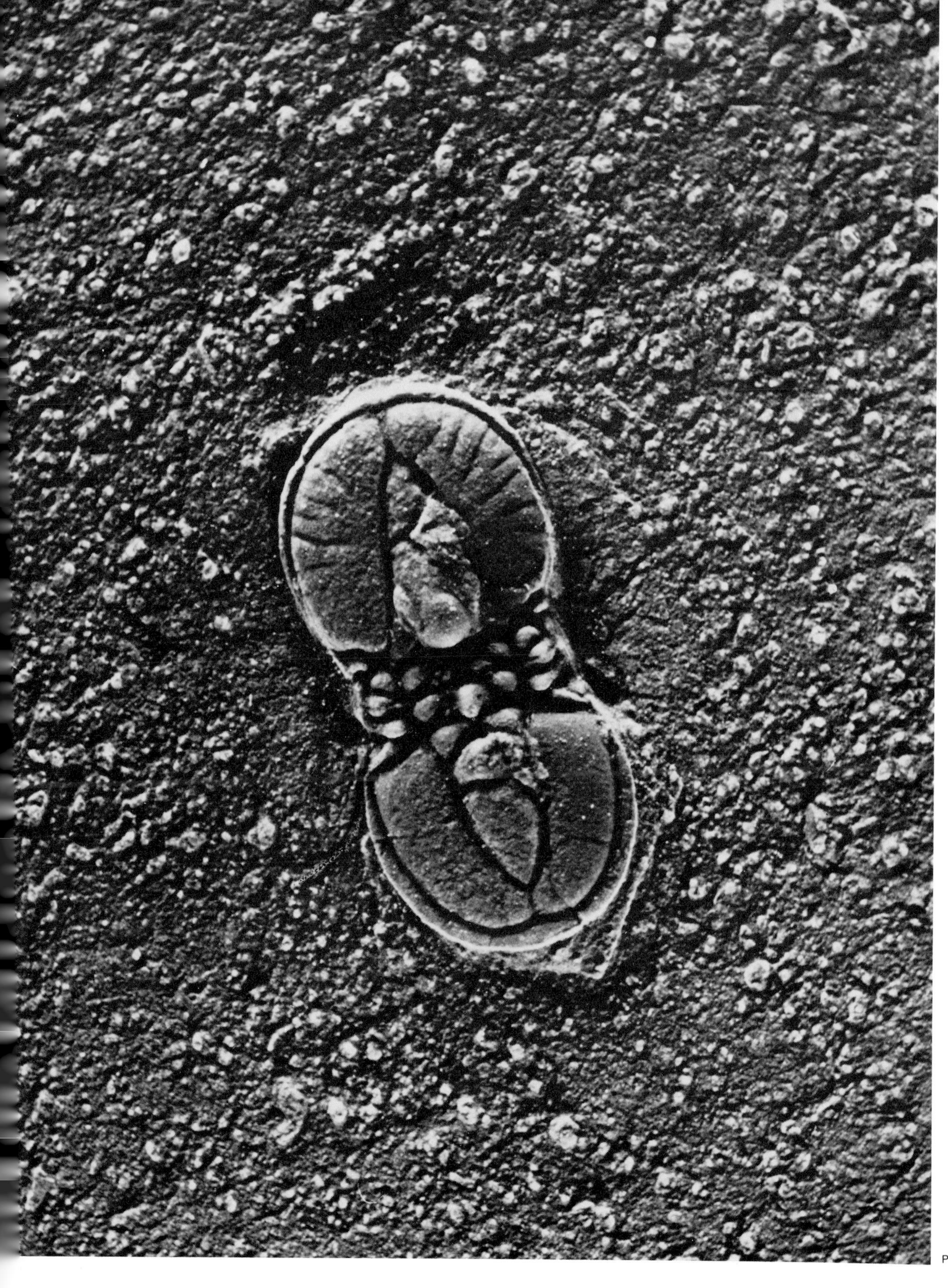

Plate 141. AGCAM 1. PTYCHAGNOSTUS Jaekel 1909. *Ptychagnostus richmondensis* (Walcott) (x18). Marjum fm., M. Cambrian, mouth of Antelope Springs Canyon, Millard Co., Utah. (Collected by A. Fawcett. RLS coll., id.) Specimen whitened with magnesium oxide. The radiating ornamentation of the cephalon is attributed to alimentary diverticula.

Pl. 141

Plate 142. AGCAM 2. *Ptychagnostus richmondensis* (Walcott) (x6.6) Wheeler shale fm., M. Cambrian, Antelope Springs, Millard Co., Utah. (Collected by A. Fawcett. RLS coll., id.) Several specimens show axial spine on second thoracic segment. The surface features emphasized for AGCAM 1 are very difficult to observe without whitening.

Pl. 142

***Family:* Diplagnostidae Whitehouse 1936**

Plate 143. AGCAM 3. BALTAGNOSTUS Lochman, 1944. *Baltagnostus eurypyx* Robison (x18). Wheeler shale, M. Cambrian, Antelope Springs, Millard Co., Utah. (Collected by A. Fawcett. RLS coll., id.) Note posterolateral marginal spines on pygidium.

Pl. 143

***Family:* Spinagnostidae Howell, 1935**

Plate 144. AGCAM 4. PERONOPSIS Corda, 1847. *Peronopsis interstricta* (White) (x19). Wheeler shale, M. Cambrian, Antelope Springs, Millard Co., Utah. (RLS coll., id.) A much enlarged view, printed from a color slide.

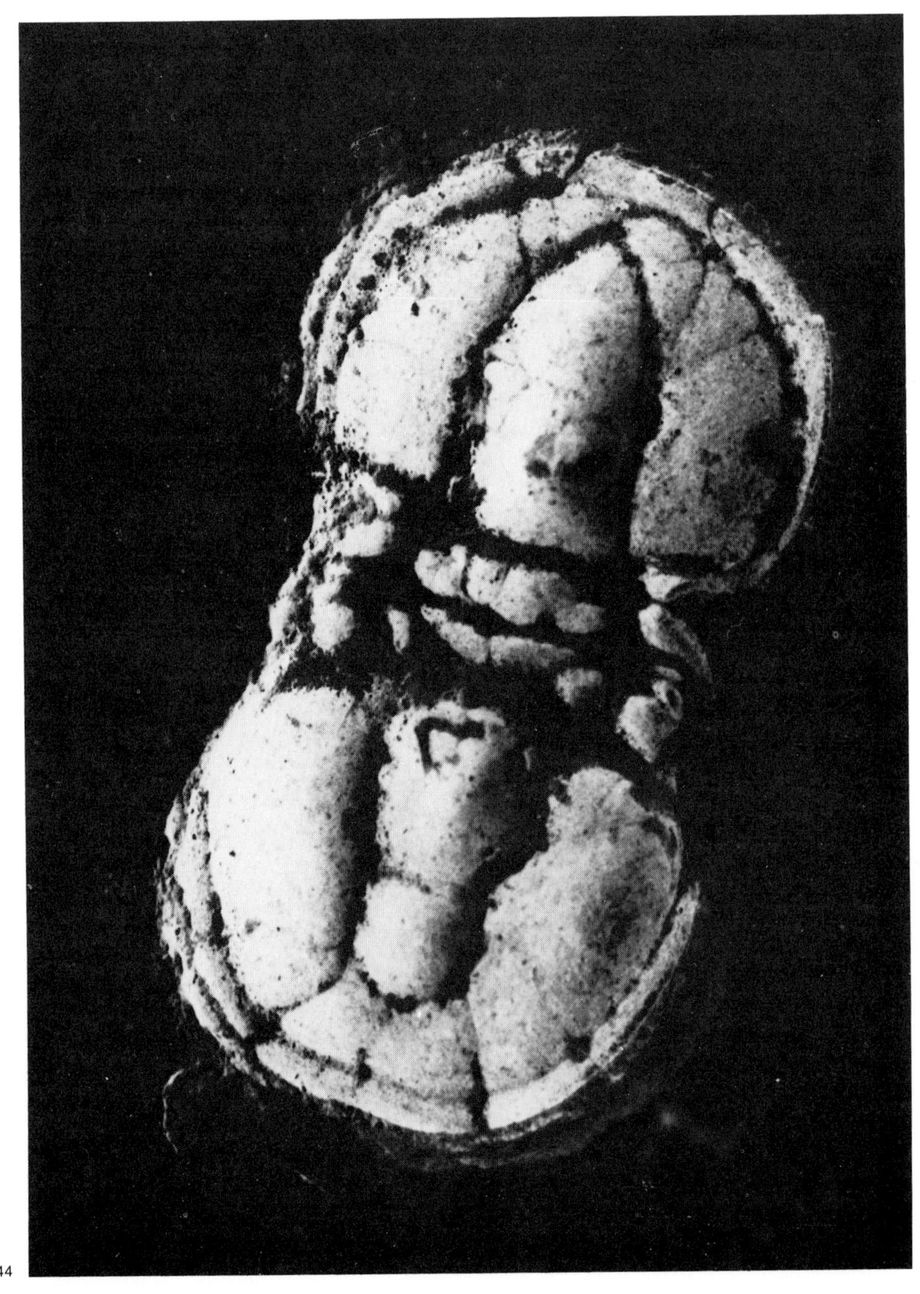

Pl. 144

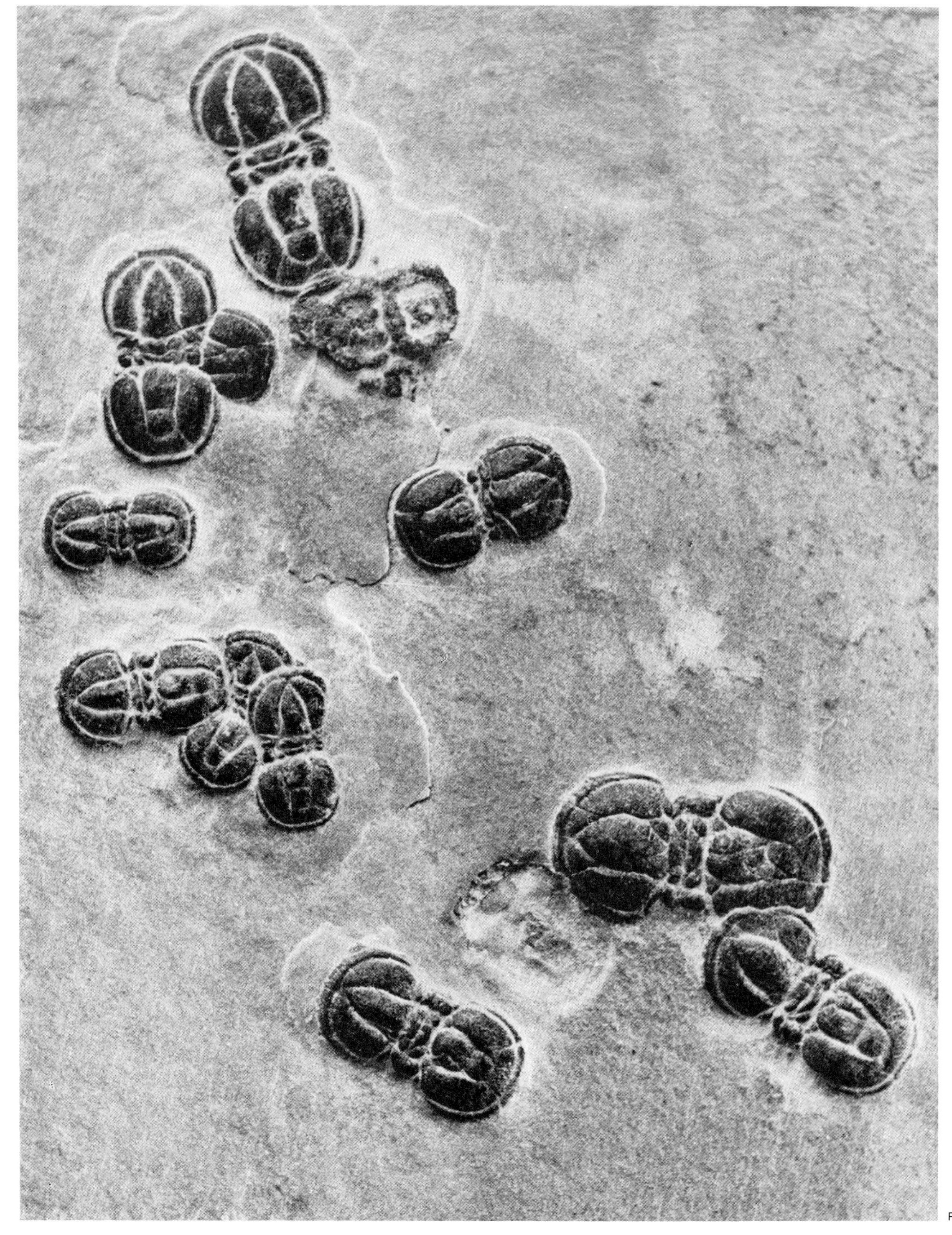

Plate 145. AGCAM 5. *Peronopsis interstricta* (White) (x7.2). Same origin as AGCAM 4. (Collected by A. Fawcett. RLS coll., id.) Assemblage of many individuals.

Pl. 145

Appendix A

Latest Memorabilia

While this volume was being composed, a number of new photographs began accumulating in the author's files. Some were contributed by paleontologist friends—alas, too late to be included at the proper location in the main body of the book. Others originated from recent findings by the author, in particular from his diggings in the Middle Cambrian *Paradoxides* beds at Manuels, Newfoundland. In view of the exceptional content of this additional photographic record, this appendix was added in proofs.

Stürmer's Trilobites

As mentioned in section 3.2, two recent studies incorporating x-ray techniques have shed much new light on the internal trilobite anatomy. One of these, by J. Cisne (1973), has already been dealt with extensively, both in section 3.2 and in the Atlas (sect. 4.2). The other important study referred to—that of Stürmer and Bergström (1973), preceded by the spectacular observations by W. Stürmer (1970)—was not illustrated in the Atlas or the text, since at the time of writing the author had not as yet succeeded in obtaining some of the original photographs for inclusion in this work. This omission can now be corrected: Prof. Stürmer's pictures arrived and some are shown here.

A slate quarry in West Germany is the depository of a paleontological treasure. The slate is the Lower Devonian Hunsrück shale, 1000 meters thick, containing a pyritized fossil fauna retaining the record of the soft parts of a variety of marine animals. Complete trilobites are found in the shale, with all appendages preserved, much as in the Utica shale of Rome, New York; sudden burial in mud, in a favorable chemical environment, may have been responsible for the exceptional preservation in both occurrences. With modern x-ray techniques, Stürmer was able to give a new description of old fossils, uncovering surprising details which had escaped previous investigations.

The classical example is provided by the comparison between an ordinary photograph of a specimen of *Phacops* (plate A1a) and its radiograph (plate A1b). The latter made news immediately (Stürmer 1970): gracefully floating with the eerie appearance of a luminescent deep sea creature, *Phacops* sp. WS 295 acquired a new dimension in contrast with its previous image in stone. The structure of the appendages, the exitic spines, the digestive tract, are sharply delineated; it should be noticed that were the radiograph of Plate A1b that of a living trilobite, the contrast would not be as good as for the pyritized fossil. The mineralization of the soft parts in the latter is actually acting as a heavy element staining of the specimen, which enhances radiographic contrast. Another beautiful radiograph, also of *Phacops* sp. is shown in Plate A2. Here the compound, schizochroal eyes of the trilobite are particularly well portrayed, and one can observe a structure of filaments apparently leading from the visual surface toward points symmetrically located on the cephalic interior. Such structure has been observed in many other examples (Stürmer and Bergström 1973), leading to the suggestion that the filaments represent crystalline fibers, much as those in the ommatidia of groups of modern arthropods. This interpretation is still the subject of some discussion, however. On the one hand, it is argued that fiber optics coupled to phacopid lenses would seem to defeat the purpose of the sophisticated design of the latter. On the other, a recent discussion (Clarkson and Levi-Setti 1974) of the visual function of schizochroal eyes proposes that these may actually be regarded as an aggregate of individual eyes rather than the mosaic-forming device common to other arthropods. In the opinion of this writer, then, the remarkable filaments in the *Phacops* eyes exhibited in Stürmer's radiographs, if positively associated with the lens structure, could well represent nerve bundles leading from individual retinal receptors to the trilobite's brain. An additional difficulty in both the fiber optics and nerve bundle hypothesis, however, remains: the filaments lead to an unlikely location for the visual center of the brain, posterior to the trilobite's esophagus. Clearly the matter is still unsettled, and further study of the Hunsrück material may provide information essential to our understanding of the function of these primordial, but not primitive, visual organs.

The Giant Trilobites of Newfoundland

Lured by vivid descriptions and by samples collected by A. M. Ziegler during his field trip to Manuels, Newfoundland, I spent a few weeks of summer vacation in 1974 quarrying in the Middle Cambrian *Paradoxides* beds in the gorge of the Manuels River. The encounter with the fauna of giant *Paradoxides* was indeed a memorable experience. More remarkable was the realization that the nature of the fossil fauna and the preservation of the trilobites rivaled the most famous comparable localities, such as Bohemia.

Lacking the means to visit the latter, I found that Manuels was in no way a cheap surrogate. In fact, the Avalon peninsula, to which Manuels belonged, was once part of the European continental plate, prior to the

drift which created the Atlantic Ocean. When the North American continent split away from Eurasia, it took along a bite of old Europe: very old indeed, since it carries rocks of Precambrian, Cambrian, and Ordovician age. The same marine animals which were buried and fossilized along a common shoreline extending from Sweden to England, and further south to a primordial Avalon, can thus be found in rocks of the same age on the other side of the Atlantic. Indeed, the Middle Cambrian *Paradoxides* beds at Manuels contain some of the trilobites found in Sweden and England.
But are they indeed the same? The work of many paleontologists, reviewed most recently by Hutchinson (1961), has seemed to indicate a close correspondence of the three faunal zones of the Middle Cambrian of the Atlantic province (in ascending order, the *Paradoxides bennetti, Paradoxides hicksi,* and *Paradoxides davidis* zones). All three have been found at Manuels (Howell 1925). In Scandinavia a further faunal zone, above the *P. davidis,* is recognized: that of *Paradoxides forchhammeri.* In Eastern Newfoundland, however, the trilobites associated with the *P. forchhammeri* zone had been found (Hutchinson 1961), but not *Paradoxides forchhammeri* itself. Where was the missing trilobite?

In my sequential quarrying of the shale beds at Manuels, I started near the top of the *Paradoxides davidis* zone. Naturally, I did not expect to find an answer to the puzzle of the missing trilobite: it was the wrong stratigraphic location for such a find. The surprise came when, after six feet of layers filled with *P. davidis,* a different kind of Paradoxides appeared, completely replacing the former. These giant trilobites, up to one foot long and colored in bright yellow, orange, and red (the colors of iron oxides coating the carapace), were too large for the average size slab to contain. It was a scramble to extract the adjoining slabs of shale and then to restore the broken specimens to their former integrity. The full realization of what the different trilobite was occurred only when, back home, I could compare my findings with the reconstructions and photographs of the *Paradoxides* of England and Scandinavia (Angelin 1851–78; Brögger 1878; Lake 1935). Gradually the conviction grew on me that I had stumbled upon *Paradoxides forchhammeri* Angelin, the missing trilobite. But it had occurred in the wrong place! This demanded a new trip to Manuels.

This time I joined forces with my friend Jan Bergström, who was visiting at Memorial University of Newfoundland. Beating the rainy season, we surveyed the Manuels beds together, this time with a yardstick. We located the very top of the *P. davidis* zone, went down to a depth of eight feet to find again my anomalous bed, and extracted more beautiful *Paradoxides forchhammeri.* This time we ascertained that the layer containing this trilobite was only about sixteen inches thick and that *P. davidis* reappeared beneath it, after some thickness of barren strata. Our findings will be described in a joint paleontological publication as soon as Jan Bergström has had a chance to compare the Newfoundland variety with the Scandinavian type specimens. The latter, as one can gather from the published record, consists mostly of trilobite fragments; we now have several complete specimens from Manuels and are in the position to give a reliable description of the species.

The previous puzzle now seems to be replaced by another one. *Paradoxides forchhammeri* in Newfoundland is located *within* the *P. davidis* zone, instead of occupying a stratigraphic position *above Paradoxides davidis,* as in Scandinavia. The two species appear to be very closely related and could have alternated locally as competing populations.

Specimens of what we believe represents *Paradoxides forchhammeri* Angelin are shown in plates A3 through A7. This species, according to our evidence, consistently is comprised of twenty thoracic segments and is characterized by a trapezoidal pygidium widening posteriorly to a spatula-like shape. The pleural spines are long and falcate. In contrast, *Paradoxides davidis* Salter, parts of which are shown for comparison in plates A8 and A9, has nineteen thoracic segments and a subrectangular, slightly tapered pygidium. The last pair of pleurae is disproportionately prolonged into saber-shaped spines extending almost a thoracic length beyond the pygidium (plate A10). Furthermore, the anterior pleural spines are shorter and more triangular than in *P. forchhammeri.* On the whole, *P. davidis* Salter is a very different beast, although the cranidium and hypostome are very similar in the two species. This may account for *P. forchhammeri* having been missed in previous investigations at Manuels. The experience also points to the value of diagnoses based on populations of complete individuals rather than on fragments of individual specimens.

The earliest description of a trilobite was given by Karl von Linné (Linnaeus 1753) of a specimen 27 cm long which he found in the Tessin Museum in Stockholm and which he named *Entomolithes paradoxus* (fig. A1, as redrawn by Angelin 1878). When Brongniart coined the name *Paradoxides* for the genus (Brongniart 1822), he argued that the Linnaeus figure, which

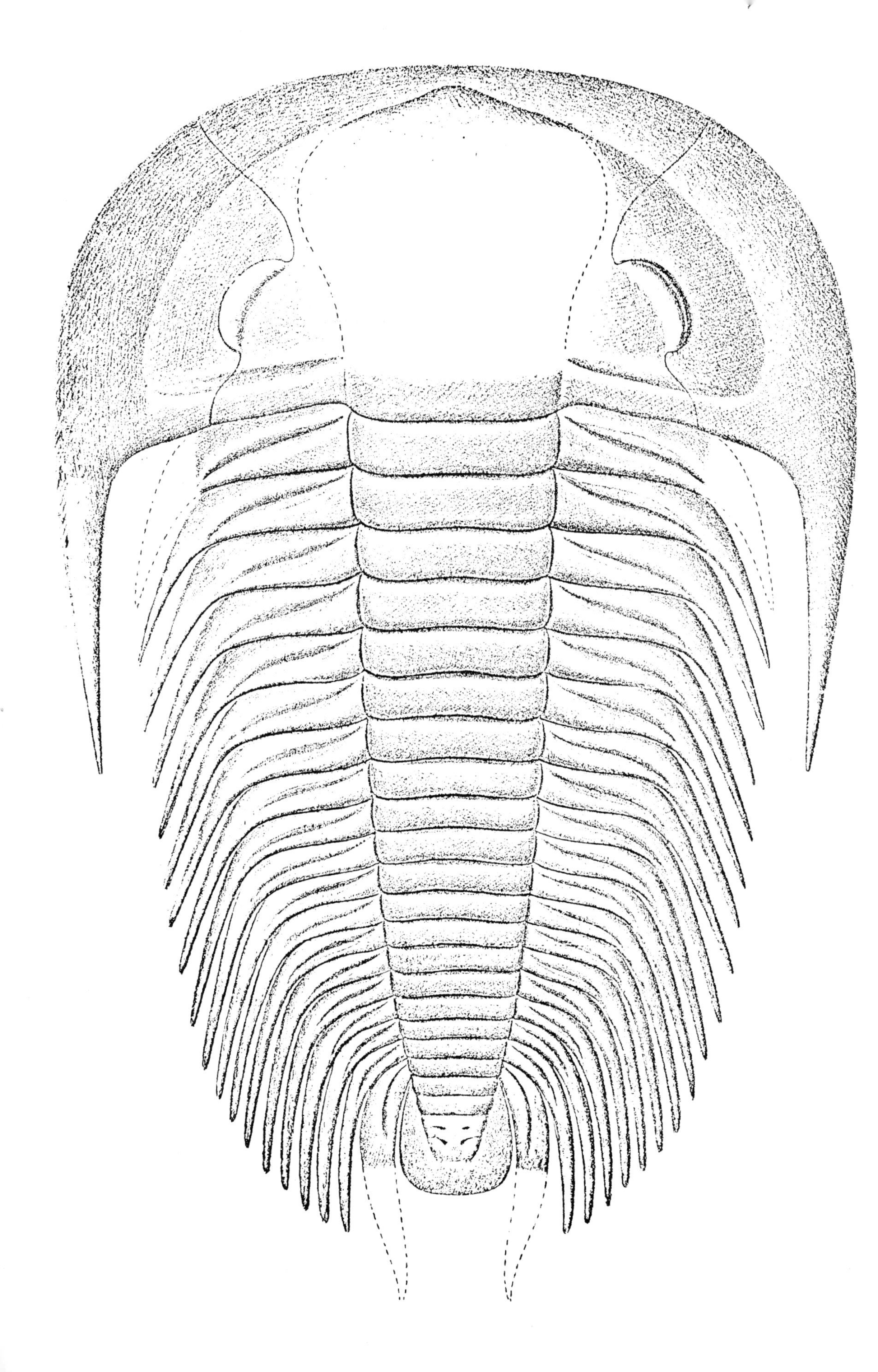

Fig. A1. Angelin's reconstruction of *Entomolithus paradoxus* Linnaeus (x 0.8) (Angelin 1878). Linnaeus noticed this unique specimen in the Tessin Museum and gave its first description in 1753 (Linnaeus 1753). According to Linnaeus, this trilobite originates from the alum shales at Oltorp, Västergötland, Sweden. Although the specimen was later assigned to the common species *Paradoxides paradoxissimus* Wahlenberg from the same locality, Angelin sets Linnaeus' specimen aside as a distinct species, other examples of which were not found after Linnaeus's.

he defines "pessima," must have represented a common Scandinavian species (although usually much smaller than Linnaeus') now called *Paradoxides paradoxissimus* (Wahlenberg). Similar inference was later made by several other authors, mostly on the basis of the original drawing, which is indeed very poor. Angelin, (Angelin 1878), however, remarked that Linnaeus's specimen was unique and that no other examples of the species had been found. As far as we can tell at this point, Angelin's reconstruction of Linnaeus's trilobite bears a remarkable resemblance to our new trilobite of Newfoundland, certainly more so than to any other *Paradoxides.* Could it be that we have rediscovered Linnaeus's giant trilobite? This we may find out, provided Linnaeus' specimen has survived two hundred twenty years of museum environment and avid collectors of Linnaean memorabilia.

A full account of the marvels at Manuels is obviously beyond the scope of this book. However, there are two unusual specimens which still must be shown. One, in plate A11, represents an uncommon occurence of *Anopolenus henrici* Salter. Although the pygidium and the free cheeks are missing, a portion of the thorax attached to the cranidium is present. The other, plate A12, shows part of the thorax and cranidium of *Centropleura* sp. This genus has never before been positively identified in Newfoundland. The specimen shown has very short pleurae and may represent a new species.

A Note on Manuels

Climbing out of the Manuels River gorge on my last trip, accompanied by my nine-year-old son Emile, I saw unmistakable signs of impending urbanization of this still-forested remnant of wilderness on the shores of Conception Bay. I left with the feeling that something must be done to preserve these amazing fossils in their natural setting at Manuels. My appeal to the authorities at Memorial University of Newfoundland found a most sympathetic response. The matter of preserving the *Paradoxides* beds at Manuels, and the small wilderness where they are exposed, is now in the hands of the Canadian Minister of Provincial Affairs for possible action.

Acknowledgment

Professor W. Stürmer's contribution of the photographs in plates A1(a) and (b) and A2 is gratefully achnowledged.

References for Appendix A

Angelin N. P. 1851–54. *Palaeontologia Scandinavica. Parts I, II—Crustacea formationis transitionis.* (Rev. ed. edited by G. Lindström. Holmiae, 1878.)

Brögger, W. C. 1878. Om Paradoxidesskifrene ved Krekling. *Nyt. Mag. Naturvit.* (Oslo) 24: 18–88.

Brongniart, A. 1822. *Histoire naturelle des crustacés fossiles, sous les rapports zoologiques et geologiques. Savoir: Les Trilobites.* Paris, F. G. Levrault.

Howell, B. F. 1925. Faunas of the Cambrian *Paradoxides* beds at Manuels, Newfoundland. *Bull. Am. Pal.* vol. 2, no. 43.

Hutchinson, R. D. 1961. Cambrian stratigraphy and trilobite faunas of southeastern Newfoundland. *Geol. Survey Canada Bull.* 88.

Lake, P. 1935. *A Monograph of the British Cambrian Trilobites.* Palaeontogr. Soc. (London). Part IX, pp. 197–224.

Linnaeus 1753. In C. G. Tessin, *Museum Tessinianum,* p. 98. Table III, fig. 1, Holmiae.

Stürmer, W. 1970. Soft parts of Cephalopods and Trilobites: Some Surprising Results of X-ray Examinations of Devonian Slates. *Science* 170: 1300–1302.

Plate A1(a). Pyritized specimen of *Phacops* sp. from the Lower Devonian Hunsrück shale, showing the appendages, exposed by mechanical preparation (x2.3); (b) X-ray photograph of the same, by W. Stürmer (WS 295). The pyritized soft parts of the trilobite, as well as the carapace, are more opaque to x-rays than the embedding shale, thus providing high contrast. Both photographs were contributed by W. Stürmer.

Pl. A1a

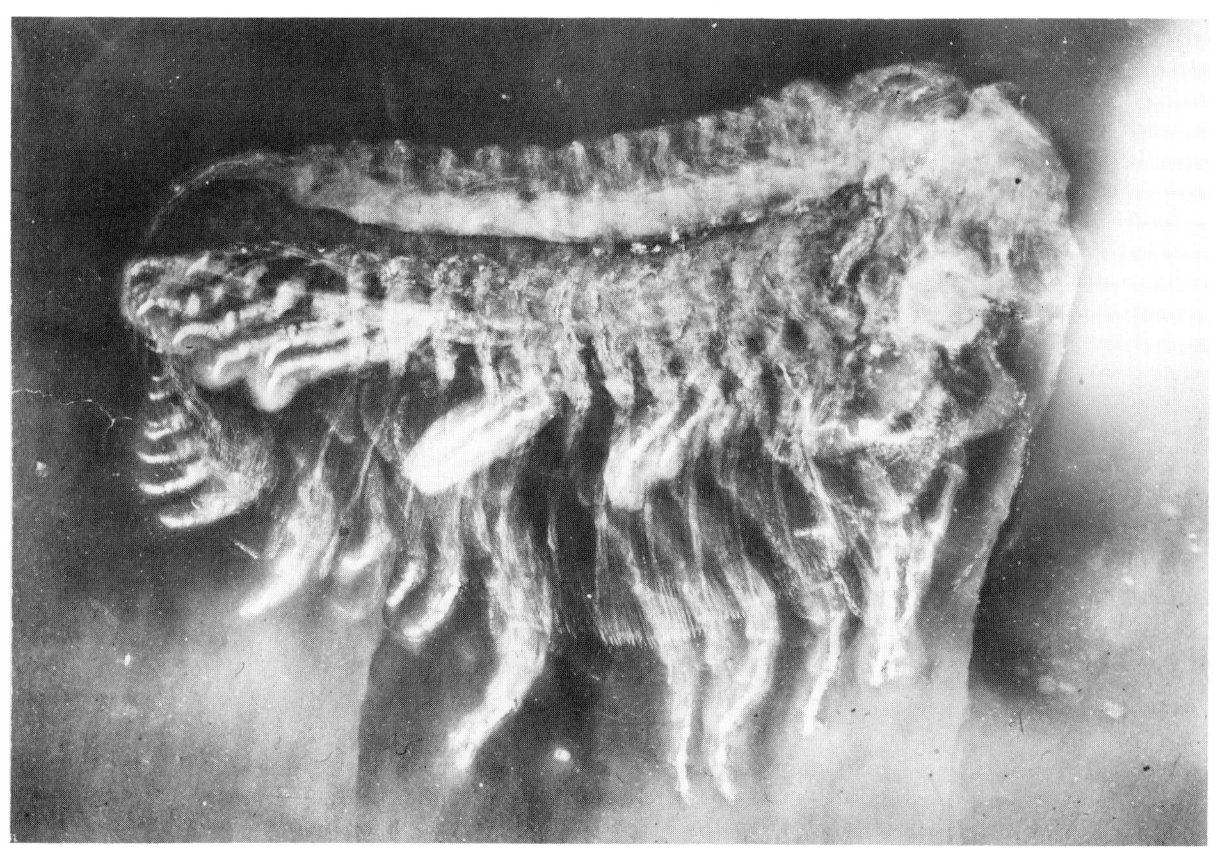

Pl. A1b

Plate A2. X-ray photograph of another specimen of *Phacops sp.* from the Hunsrück shale (WS 2617, x5). Fibers apparently associated with the compound eye structure are visible in the region of the palpebral lobes. Photograph contributed by W. Stürmer.

Pl. A2

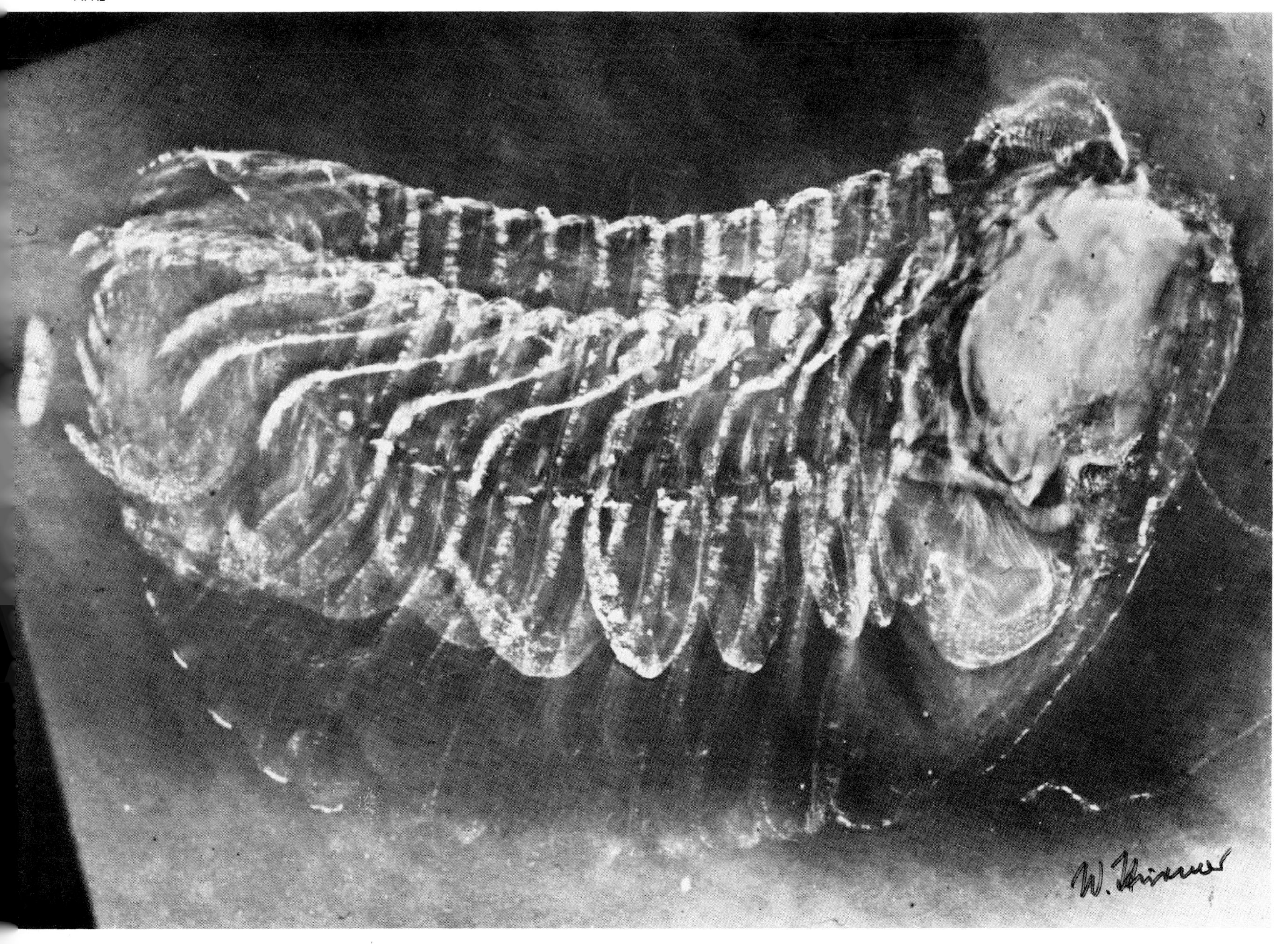

Pl. A3

Plate A3. RECAM 23. PARADOXIDES Brongniart, 1822. *Paradoxides forchhammeri* Angelin (preliminary identification) (x 1.5). Middle Cambrian, Manuels (RLS coll., id.). Negative mold. The trapezoidal spatulated pygidium and the diverging last pair of pleurae are a distinctive feature of this species, which is characteristic of the Scandinavian fauna. The cranidium and free cheeks are incomplete in this example.

Plate A4. RECAM 24. Partial view of another example of *Paradoxides forchhammeri* Angelin (preliminary identification) as in plate A3 (x 2, RLS coll., id.). Print from Kodachrome slide.

Pl. A4

Plate A5. RECAM 25. Almost complete exuviae of *Paradoxides forchhammeri* Angelin (preliminary identification) as in plate A3 (natural size, RLS coll., id.). The free cheeks, loosened in the molting process, are displaced from their original position and overlap the lateral edges of the cranidium. The genal spines of this otherwise exceptional specimen could not be recovered. Print from Kodachrome slide.

Pl. A5

Plate A6. RECAM 26. *Paradoxides forchhammeri* Angelin (preliminary identification) as in plate 5 (natural size, RLS coll., id.). Although the cranidium is missing in these exuviae, the free cheeks are positioned in their original location giving correct proportions to the overall outline of this trilobite. Print from Kodachrome slide.

Pl. A6

Plate A7. RECAM 27. Complete exuviae of *Paradoxides forchhammeri* Angelin (preliminary identification), another of the many examples of this trilobite discovered at Manuels, Newfoundland. (x 1.2, RLS coll., id.). Both free cheeks are displaced from their original setting; one is overturned. The cranidium is partly crushed, exposing the underlying hypostome, also displaced. This specimen is the prize finding of Emile Levi-Setti (by now age 9—see caption to plate 131) when he was given the chance to quarry a slab of his own in September 1974.

Pl. A7

Pl. A8

Plate A8. RECAM 28. *Paradoxides davidis* Salter (x 1.7 RLS coll., id.), Middle Cambrian, Manuels, Newfoundland. Partially disarticulated exuviae; the cranidium is missing. This species is characteristic of the *Paradoxides* of Great Britain and Sweden.

Pl. A9

Plate A9. RECAM 29. Complete thorax and part of the cranidium, exuviae of *Paradoxides davidis* Salter (x 1.6, RLS coll., id.). Origin as in plate A8. The pleural spines are only partially exposed.

Plate A10. RECAM 30. Pygidial region of *Paradoxides davidis* Salter, origin as in plate 8 (x 2.1, RLS coll., id.). This detail illustrates one of the most outstanding morphological differences between this species and that of Plates A3 through A7, attributed to *P. forchhammeri* Angelin. Note the tapered pygidium and the saber-like, x-shaped last pair of pleural spines.

Pl. A10

Plate A11. RECAM 31.
ANOPOLENUS Salter 1864.
Anopolenus henrici Salter (x 2.3).
Middle Cambrian *Paradoxides davidis* zone at Manuels, Newfoundland (RLS coll., id.). This is a trilobite characteristic of the British Centropleurinae, peculiar in the truncated appearance of the pleurae. The occurrence of part of the thorax still articulated to the cranidium is uncommon. The free cheeks, part of the thorax, and pygidium are missing. Print from Kodachrome slide.

Pl. A11

Plate A12. RECAM 32. CENTROPLEURA Angelin, 1854. *Centropleura sp.* (x 2.6). Middle Cambrian *Paradoxides hicksi* zone at Manuels, Newfoundland (RLS coll., id. by J. Bergström). Part of the thorax and cranidium. This is the first example of this genus positively identified in Newfoundland and most probably is a new species. Specimen collected by Susanna Ginali.

Pl. A12

A Farewell Photograph

Olenellid trilobites have undoubtedly a particular appeal. This could be due to their elegant shape or to their being among the most ancient trilobites. Thus the desire to include one more picture of this kind is compelling. Plate A13 represents an adult individual of *Olenellus clarki* (Resser) from the Lower Cambrian of San Bernardino County, California. Aside from the attractive posture and outline of this trilobite, a significant feature is also present. The diminutive flat pygidium and the extension of the rachis beyond the thorax proper to form the so-called opisthothorax are visible here. These are distinctive characteristics of the Olenellinae which were not exhibited in the pictures of section 4.1.

Plate A13. OLCAM 6. OLENELLUS Billings, 1861. *Olenellus clarki* (Resser) (x 3.1). Lower Cambrian, San Bernardino County, California (RLS coll., id.). Negative mold. The discussion in the caption to plate 39 is relevent here. Note the exposure of the opisthothorax in this graceful example; the presence of this extension of the rachis is characteristic of the Olenellinae.

Pl. A13

Appendix B: Photographic Techniques and Specimen Preparation

Photographic Techniques

Very little can be added to the description of the physical principles involved and the technique of fossil photography given by Franco Rasetti in the *Handbook of Paleontological Techniques* (Kummel and Raup 1965). As a physicist, I have appreciated Rasetti's suggestions tremendously and have endeavoured to translate them into practice as much as possible.

Although it is not generally recommended for high resolution work, 35mm. film was used for all of the photographs. The use of slow, fine-grained film and careful development procedure has, however, enabled enlargements to 8" × 10" print size without loss of quality. For this purpose a now irreplaceable negative film was used—ADOX KB 14, 20 ASA. Fortunately, most of the photographs were taken before this film was discontinued. Film based on the same patent is now made in Yugoslavia with the designation EFKE KB 14, but it is unavailable in the U.S.A. The film was developed in AGFA Rodinal developer, diluted 1:100, for 20 minutes at 20°C. Microscopic examination of the developed image in ADOX KB 14 shows a grain size and lack of graininess (coalescence of groups of grains) without rival in any other presently available negative material.

I seem to have found my best solutions to the problems of fossil photography in the use of materials and equipment now discontinued. Among the latter is an item which is responsible for good focusing in the pictures presented here, a Leitz device called Reprovit. It consists of a sliding mount, carrying the camera body (a Leica) and a ground glass placed at the level of the film plane in the camera. Beneath the sliding mount, the lens is usually carried by a focusing extension or, as modified by the author, a bellows controlled by rack and pinion motion. Focusing is achieved by projecting the image of a flat marker, encased in the ground glass, onto the surface of a specimen, as if the device were a photographic enlarger. By doing so, the field of view seen by the lens is also illuminated, and the specimen can be properly centered or positioned. The lens stop is then closed to the desired setting, the projector removed and the camera made to slide into position above the lens. The focal plane shutter is open and the exposure takes place by switching the illuminating lamps on and off. Depending on the size of the specimen and depth of field required, different lenses were used, usually stopped to the largest f: number. Among the lenses used are Leitz Elmar 50mm. (f: 16), Schneider Componon (f: 22), and Schneider Symmar 135mm. (f: 45). The latter two are projection lenses, originating from the author's bubble chamber film scanning projectors.

For printing, a Leitz enlarger with 50mm. Elmar lens was used. All enlargements were printed on Agfa Brovira paper, mostly extra-hard, developed in Kodak Dektol developer diluted 1:2.

Specimen illumination was achieved generally with two high-intensity desk miniature lamps, at times with a parallel beam of light from a microscope illuminator, and occasionally with a ring flourescent lamp.

As described in Rasetti's article, the immersion of specimens in xylene often enhances the contrast between the specimen and matrix. This technique was employed to particular advantage when mineralized or pigmented details were embedded in surface incrustations of calcite or quartz. The optical contact between these minerals and the xylene, having approximately the same refractive index, enables light to penetrate unreflected and unrefracted through the surface layers, revealing the underlying structure of the fossil. Immersion in xylene yielded quite spectacular results also in enhancing the contrast of the pyritized appendages of the Utica shale trilobites. In several instances, when color contrast existed between specimen and matrix, it was found worthwhile to take a color photograph on a Kodachrome slide and then print directly from the slide. The resulting print is a negative, but this is irrelevant in fossil photography. The high contrast factor of Kodachrome (not Ektachrome) yields extremely sharp prints.

Valuable sources of further technical information on the photography of small objects in general are two Kodak technical publication manuals—N 12A on close-up photography, and N 12B on photomacrography.

Specimen Preparation

The preparation of the specimens for photography has often been a painstaking operation, usually carried out under a stereoscopic microscope. Some of the methods adopted followed the advice of Rasetti and Palmer in the *Handbook of Paleontological Techniques* (Kummel and Raup 1965). Different procedures have been employed in exposing trilobites preserved in shale from those preserved in, say, limestone. Soft shale—for example, the Silica shale—disintegrates in water, and so does the trilobite. Here the matrix is easily removed to expose the trilobite details by using a variety of scraping tools and soft bristle brushes. Metal brushes are to be avoided at all times. The use of varnish or shellac to waterproof or consolidate the exoskeleton has been avoided in order not to cause unwanted reflections. Repairs of flaked parts have been successfully obtained by using Duco cement diluted in Amyl Acetate.

(The suggestion is from Rasetti and Palmer.) This diluted glue dries without leaving a glossy surface. Final cleaning of the trilobite was obtained by using a moist cloth. A somewhat harder shale is the Waldron shale or Wheeler shale. Here prolonged soaking in a solution of Quaternary O (Geigy Industrial Chemicals) is the best preparation. Caution should be taken to avoid any procedure which could scratch the calcite of the carapace. Unless the specimen is whitened, chisel marks are very disturbing to good photography. In general, care also has been taken to present the matrix in a fairly natural condition.

The hardest preparation is that of trilobites in limestone. Only rarely, as in specimens from Grafton, Illinois, does the separation of trilobite and matrix obtain naturally. More often the trilobite can only be exposed by resorting to rather arduous chiseling and grinding and to the aid of vibro-tools. In these cases the matrix surface is usually left scarred by white marks. In such cases careful illumination has been used in order to form shadows on the unwanted scratches.

The trilobites presented are, in many instances, the only survivors of unsuccessful attempts to prepare a much larger initial sample. For obvious reasons, the specimens borrowed from museums have been left untouched.

It is standard procedure in paleontology to whiten fossils for photography. This approach has been followed here only in a few instances, by exposing the specimen to magnesium oxide vapors from a burning magnesium ribbon (Rasetti 1947).

In conclusion, this book contains a spectrum of both orthodox and unorthodox approaches to the photography of fossils. These notes contain no implication whatsoever that any of the methods here employed are preferable to the accepted standards in paleontology. My personal tastes and the means at my disposal have been the only guidelines here.

References

A. Technical literature specifically referred to in the text

Beecher, C. E. 1897. Outline of a natural classification of the trilobites. *Am. Jour. Sci,* ser. 4, vol. 3, no. 13, pp. 89–106, 181–207, pl. 3.

Bergström, J. 1973a. Organization, life, and systematics of trilobites. *Fossils and strata,* no. 2. Oslo: Universitetsforlaget Oslo.

______. 1973b. Classification of olenellid trilobites and some Balto-scandian species. *Norsk Geolog. Tidsskr.* 53: 283–314.

Cisne, J. 1973. The anatomy of *Triarthrus eatoni* and its bearing upon the phylogeny of the Trilobita and Arthropoda. Ph.D. dissertation, University of Chicago.

Clarkson, E. N. K. 1966. Schizochroal eyes and vision in some phacopid trilobites. *Palaeontology* 9:464–487.

______. 1968. Structure of the eye of *Crozonaspis struvei. Senckenbergiana Lethaea* 49:383–93.

______. 1969. On the schizochroal eyes of three species of *Reedops* (Trilobita: Phacopidae) from the Lower Devonian of Bohemia. *Royal Soc. of Edinburgh Trans.* 68:183–205.

______. 1973a. Private communication.

______. 1973b. The eyes of *Asaphus raniceps* Dalman (Trilobita). Palaeontology 16:425–44.

______. 1973c. Morphology and evolution of the eye of Upper Cambrian Olenidae (Trilobita). *Palaeontology* 16:735–63.

Dalman, J. W. 1927. Om Palaeaderna eller de så Kallade Trilobiterna. *K. Svenska Vetensk. Akad, Handl.* 1:226–94.

Descartes, R. 1637. *La géometrie.* Originally published in 1637. Reproduced in *Oeuvres de Descartes,* vol. 6, edited by C. Adam and P. Tannery. Paris: Librairie Philosophique, J. Vrin.

Easton, W. H. 1960. *Invertebrate paleontology.* New York: Harper and Bros.

Eldredge, N. 1972. Systematics and evolution of *Phacops rana* (Green, 1832) and *Phacops iowensis* Delo, 1935 (Trilobita) from the Middle Devonian of North America. *Bull. Am. Mus. Nat. Hist.* 147:49–113.

Emmrich, H. F. 1839. Ueber die Trilobiten. *Neues-Jahrb. f. Mineral.* (Stuttgart), 1845, pp. 18–52.

Exner, S. 1891. *Die Physiologie der facettierten Augen von Krebsen und Insekten.* Leipzig and Vienna: Verlag Fr. Deuticke.

Fritz, W. H. 1972. Lower Cambrian trilobites from the Sekwi Formation type section, Mackenzie Mountains, Northwestern Canada. *Geol. Survey of Canada Bull.* 212.

Hartline, H. K. 1969. Visual receptors and retinal interaction. *Science* 164:270–78.

Henningsmoen, G. 1951. Remarks on the classification of trilobites. *Norsk. Geol. Tidsskr.* 29:174–217.

Höglund, G. 1965. Pigment migration and retinular sensitivity. In *The functional organization of the compound eye,* edited by C. G. Bernhard. Elmsford, N.Y.: Pergamon Press.

Horridge, G. A. 1969. The eye of the firefly *Photuris. Proc. Roy. Soc. B.* 171:445–63.

Hupé, P. 1953. Classification des trilobites. *Ann. Paleont.* 39 (1953):59–168 and 41 (1955): 91–325.

Huygens, C. 1960. *Treatise on light.* Original published in French, 1690. English translation by S. P. Thompson, 1912. Chicago: University of Chicago Press.

Kummel, B. 1970. *History of the earth: An introduction to historical geology.* 2d ed. San Francisco: W. H. Freeman and Co.

Kummel, B., and Raup, D., eds. 1965. *Handbook of paleontological techniques.* San Francisco: W. J. Freeman and Co.

Levi-Setti, R.; Park, D; and Winston, R. 1973. The corneal cones of *Limulus* as optimized light concentrators. *Nature,* in the press.

Lindström, G. 1901. Researches on the visual organs of the trilobites. *K. Svenska Vetensk. Akad. Handl.* 34: 1–89.

Milne-Edwards, H. 1840. *Histoire naturelle des crustacés.* Paris.

Moore, R. C., ed. 1959. *Treatise on invertebrate paleontology.* Part O. Lawrence, Kans.: Geological Society of America and University of Kansas Press. This is the basic and most comprehensive reference on trilobites.

Müller, J. 1826. Zur vergleichenden Physiologie des Gesichtsinnes. Leipzig: Cnobloch.

Öpik, A. A. 1961. Alimentary caeca of agnostids and other trilobites. *Palaeontology* 3:410–38.

Rasetti, F. 1947. Notes on techniques in invertebrate paleontology. *J. Paleontology* 21:397–98.

Raymond, P. E. 1939. *Prehistoric life.* Princeton: Harvard University Press.

Richter, R. 1953. Crustacea (Paläontologie). In *Handwörterbuch der Naturwissenschaften.*2d ed., vol. 2, no. 840–64.

Roy, S. K. 1933. A new Devonian trilobite from southern Illinois. *Fieldiana: Geology* 6:67–82.

Shrock, R. R., and Twenhofel, W. H. 1953. *Principles of invertebrate paleontology.* New York: McGraw-Hill, Inc.

Snodgrass, R. E. 1952. *A textbook of arthropod anatomy.* Ithaca: Comstock Publishing Assoc.

Stürmer, W., and Bergström, J. 1973. New discoveries on trilobites by x-rays. *Paläont. Z.* 47:104–141.

Towe, K. M. 1973. Trilobite eyes: calcified lenses *in vivo. Science* 179:1007–9.

B. Relevant textbooks on invertebrate paleontology and zoology which have been the source of much information incorporated in this book.

Easton, W. H. 1960. *Invertebrate paleontology.* New York: Harper and Bros.

Hupé, P. 1953. Classe des trilobites. In J. Piveteau, ed., *Traité de paleontologie.* vol. 3. Paris: Masson et Co.

Kummel, B. 1970. *History of the earth: An introduction to historical geology.* 2d ed. San Francisco: W. H. Freeman and Co.

Rockstein, M. 1964. *The physiology of insecta.* New York: Academic Press.

Schrock, R. R., and Twenhofel, W. H. 1953. *Principles of invertebrate paleontology.* New York: McGraw Hill, Inc.

Waterman, T. H. 1961. *The physiology of crustacea.* New York: Academic Press.

C. On fossils in general, mostly nontechnical.

MacFall, R. P., and Wollin, J. 1972. *Fossils for amateurs.* New York: Van Nostrand. Broad coverage of fossil collecting and preparation. (Several photographs originate from R. Levi-Setti, apparently without knowledge on the part of the authors, and the fossils represented are unfortunately misidentified.)

Müller, A. H. 1962. *Aus Jahrmillionen: Tiere der Vorzeit.* Jena: G. Fisher Verlag. A beautiful collection of large format photographs of invertebrate and vertebrate fossils.

Pinna, G. 1972. *The dawn of life.* New York: World Publishing. Invertebrate fossils in captivating color photographs. (Several trilobites are incorrectly identified.)

Ransom, J. E. 1964. *Fossils in America.* New York: Harper and Row. A useful introduction and field guide to fossil collecting.

D. Technical literature consulted in the preparation of this work

These are only a few of a much larger body of references, most of which are listed in Moore (1959), which have been perused.

Bright, R. C. 1959. A paleoecologic and biometric study of the Middle Cambrian trilobite *Elrathia kingii* (Meek). *J. Paleontology* 33:83–98.

Hall, J. 1879. The fauna of the Niagara group, in Central Indiana. *28th Annual Report of the New York State Museum of Natural History, pp.* 99–210.

Merriam, C. W., and Palmer, A. R. 1964. *Cambrian rocks of the Pioche mining district, Nevada.* Geol. Surv. Prof. Paper 469. Washington, D.C.: U.S. Gov. Printing Office.

Rasetti, F. 1951. Middle Cambrian stratigraphy and faunas of the Canadian Rocky Mountains. *Smithsonian Misc. Coll.* 116, no. 5., pp. 1–277.

Resser, C. E. 1928. Cambrian fossils from the Mohave Desert. *Smithsonian Misc. Coll.* 81, no. 2, pp. 1–15.

______. 1939. The Spence shale and its fauna. *Smithsonian Misc. Coll.* 97. no. 12, pp. 1–29.

Robison, R. A. 1964. Late Middle Cambrian faunas from western Utah. *J. Paleontology* 38:510–66.

______. 1971 Additional Middle Cambrian trilobites from the Wheeler shale of Utah. *J. Paleontology* 45:794–804.

Stumm, E. C., and Kauffman, E. G. 1958. Calymenid trilobites from the Ordovician rocks of Michigan. *J. Paleontology* 32: 943–60.

Weller, S. 1907. The paleontology of the Niagaran limestone in the Chicago area: The Trilobita. *Chicago Acad. Sci. Bull. IV,* part II of *Natural History Survey,* 20 May 1907.